Metal Nanoparticles as Catalysts for Green Applications

Metal Nanoparticles as Catalysts for Green Applications

Editors

**Michela Signoretto
Federica Menegazzo**

MDPI • Basel • Beijing • Wuhan • Barcelona • Belgrade • Manchester • Tokyo • Cluj • Tianjin

Editors
Michela Signoretto
Department of Molecular
Sciences and Nanosystems
Ca' Foscari University of
Venice
Venice
Italy

Federica Menegazzo
Department of Molecular
Sciences and Nanosystems
Ca' Foscari University of
Venice
Venice
Italy

Editorial Office
MDPI
St. Alban-Anlage 66
4052 Basel, Switzerland

This is a reprint of articles from the Special Issue published online in the open access journal *Processes* (ISSN 2227-9717) (available at: www.mdpi.com/journal/processes/special_issues/metal_nano_catal).

For citation purposes, cite each article independently as indicated on the article page online and as indicated below:

LastName, A.A.; LastName, B.B.; LastName, C.C. Article Title. *Journal Name* **Year**, *Volume Number*, Page Range.

ISBN 978-3-0365-4474-8 (Hbk)
ISBN 978-3-0365-4473-1 (PDF)

© 2022 by the authors. Articles in this book are Open Access and distributed under the Creative Commons Attribution (CC BY) license, which allows users to download, copy and build upon published articles, as long as the author and publisher are properly credited, which ensures maximum dissemination and a wider impact of our publications.
The book as a whole is distributed by MDPI under the terms and conditions of the Creative Commons license CC BY-NC-ND.

Contents

About the Editors vii

Michela Signoretto and Federica Menegazzo
Special Issue "Metal Nanoparticles as Catalysts for Green Applications"
Reprinted from: *Processes* **2021**, *9*, 1015, doi:10.3390/pr9061015 1

Nga Tran, Yoshimitsu Uemura, Thanh Trinh and Anita Ramli
Hydrodeoxygenation of Guaiacol over Pd–Co and Pd–Fe Catalysts: Deactivation and Regeneration
Reprinted from: *Processes* **2021**, *9*, 430, doi:10.3390/pr9030430 5

Somayeh Taghavi, Elena Ghedini, Federica Menegazzo, Michela Signoretto, Delia Gazzoli and Daniela Pietrogiacomi et al.
MCM-41 Supported Co-Based Bimetallic Catalysts for Aqueous Phase Transformation of Glucose to Biochemicals
Reprinted from: *Processes* **2020**, *8*, 843, doi:10.3390/pr8070843 15

Danilo Bonincontro, Francesco Fraschetti, Claire Squarzoni, Laura Mazzocchetti, Emanuele Maccaferri and Loris Giorgini et al.
Pd/Au Based Catalyst Immobilization in Polymeric Nanofibrous Membranes via Electrospinning for the Selective Oxidation of 5-Hydroxymethylfurfural
Reprinted from: *Processes* **2020**, *8*, 45, doi:10.3390/pr8010045 31

Cristina Pizzolitto, Federica Menegazzo, Elena Ghedini, Arturo Martínez Arias, Vicente Cortés Corberán and Michela Signoretto
Microemulsion vs. Precipitation: Which Is the Best Synthesis of Nickel–Ceria Catalysts for Ethanol Steam Reforming?
Reprinted from: *Processes* **2020**, *9*, 77, doi:10.3390/pr9010077 47

Nimita Francy, Subramanian Shanthakumar, Fulvia Chiampo and Yendaluru Raja Sekhar
Remediation of Lead and Nickel Contaminated Soil Using Nanoscale Zero-Valent Iron (nZVI) Particles Synthesized Using Green Leaves: First Results
Reprinted from: *Processes* **2020**, *8*, 1453, doi:10.3390/pr8111453 61

Anis Hamza Fakeeha, Yasir Arafat, Ahmed Aidid Ibrahim, Hamid Shaikh, Hanan Atia and Ahmed Elhag Abasaeed et al.
Highly Selective Syngas/H_2 Production via Partial Oxidation of CH_4 Using (Ni, Co and Ni–Co)/ZrO_2–Al_2O_3 Catalysts: Influence of Calcination Temperature
Reprinted from: *Processes* **2019**, *7*, 141, doi:10.3390/pr7030141 73

Trinh Duy Nguyen, Oanh Kim Thi Nguyen, Thuan Van Tran, Vinh Huu Nguyen, Long Giang Bach and Nhan Viet Tran et al.
The Synthesis of N-(Pyridin-2-yl)-Benzamides from Aminopyridine and Trans-Beta-Nitrostyrene by Fe_2Ni-BDC Bimetallic Metal–Organic Frameworks
Reprinted from: *Processes* **2019**, *7*, 789, doi:10.3390/pr7110789 89

Ourmazd Dehghani, Mohammad Reza Rahimpour and Alireza Shariati
An Experimental Approach on Industrial Pd-Ag Supported -Al_2O_3 Catalyst Used in Acetylene Hydrogenation Process: Mechanism, Kinetic and Catalyst Decay
Reprinted from: *Processes* **2019**, *7*, 136, doi:10.3390/pr7030136 103

About the Editors

Michela Signoretto

Michela Signoretto is Full Professor of Industrial Chemistry at Ca'Foscari University of Venice. She leads CATMAT research group at the Department of Molecular Sciences and Nanosystems. She has co-authored nine book chapters and over 130 scientific publications on international journals in the field of heterogeneous catalysis and material sciences. Her research focuses on nanomaterials synthesis for energy and environmental applications. She is also involved in the formulation of new sustainable materials for pharmaceutical and cosmetic applications.

Federica Menegazzo

Federica Menegazzo received her Ph.D. degree in Chemical Sciences at the University of Ferrara in 2003. Currently she works as Associate Professor in Industrial Chemistry at the Department of Molecular Sciences and Nanosystems, Ca'Foscari University of Venice. Her principal scientific interests are in the field of heterogeneous catalysis with a focus on the development of innovative nanostructured catalysts and their use in industrial and sustainable chemistry. She has co-authored five book chapters and over 90 publications in peer-reviewed journals.

Editorial
Special Issue "Metal Nanoparticles as Catalysts for Green Applications"

Michela Signoretto * and Federica Menegazzo *

Department of Molecular Sciences and Nano Systems, Università Ca' Foscari Venezia, Via Torino 155, 30172 Venezia, Italy
* Correspondence: miky@unive.it (M.S.); fmenegaz@unive.it (F.M.)

Citation: Signoretto, M.; Menegazzo, F. Special Issue "Metal Nanoparticles as Catalysts for Green Applications". *Processes* **2021**, *9*, 1015. https://doi.org/10.3390/pr9061015

Received: 5 June 2021
Accepted: 7 June 2021
Published: 9 June 2021

Publisher's Note: MDPI stays neutral with regard to jurisdictional claims in published maps and institutional affiliations.

Copyright: © 2021 by the authors. Licensee MDPI, Basel, Switzerland. This article is an open access article distributed under the terms and conditions of the Creative Commons Attribution (CC BY) license (https://creativecommons.org/licenses/by/4.0/).

This Special Issue of *Processes* on "Metal Nanoparticles as Catalysts for Green Applications" collects recent works of researchers on metal nanoparticles as catalysts for green applications. All applications that deal with designing chemical products and processes that generate and use less (or preferably no) hazardous substances, by applying the principles of green chemistry, were welcome for this Special Issue. Despite the interdisciplinary nature of the different applications involved, ranging from pure chemistry to material science, from chemical engineering to physical chemistry, in this Special Issue there are common characteristics connecting the areas together, and they can be described by two words: sustainability and catalysis. We are convinced that a strategic goal of our world is the development of a sustainable society, which is one that "meets the needs of the current generation without sacrificing the ability to meet the needs of future generations" [1]. Catalysis, which represents probably the oldest application of nanotechnology, has a key role on the road to sustainability. Therefore, we are proud to contribute to the knowledge and deepening of these two concepts.

The journal *Processes* covers a wide range of materials related topics including the formulation of catalysts, process technology, and applications. Such diversity is reflected in this Special Issue through eight contributions.

We believe that the advances described by the different investigations have appreciably helped with reaching toward the target of meaningful sustainability. In fact, most of the papers deal with reactions for biomass valorization, covering a wide range of applications, which highlights the versatility of the subject matter. For instance, guaiacol has been upgraded by hydrodeoxygenation [2], glucose by aqueous phase reforming [3], 5-hydroxymethylfurfural (HMF) by selective oxidation [4], and ethanol by steam reforming [5].

The biomasses have been used also by themselves for the formulation of innovative catalysts, as in the case of green leaves for the synthesis of iron particles [6].

Another issue that has been faced up to is hydrogen production [5,7] since it is considered the future energy vector, and energy is a top concept in a sustainable vision.

On the topic of biomasses valorization, the paper by Nga Tran et al. [2] investigates the hydrodeoxygenation of guaiacol over Pd-Co and Pd-Fe catalysts supported on Al-MCM-41, with a focus on stability and regeneration, which are of great importance in bio-oil upgrading. The authors found that the bimetallic Pd-Co and Pd-Fe showed a higher yield and stability than monometallic Co and Fe, since the coke formation was reduced. In particular, the Pd-Fe catalyst presented even a higher stability and regeneration ability than the Pd-Co catalyst.

In the paper by Taghavi et al. [3], glucose was upgraded to valuable biochemicals such as fructose, levulinic acid, ethanol, and hydroxyacetone by aqueous phase transformation. Different MCM-41-supported metallic and bimetallic (Co, Co-Fe, Co-Mn, Co-Mo) catalysts, as well as different catalysts under different reaction conditions, were synthesized and characterized by numerous techniques. The authors demonstrated that reaction conditions, bimetals synergetic effects, and the amount and strength of catalyst acid sites were the key factors affecting the catalytic activity and biochemical selectivity. Best results (i.e., the highest carbon balance and the desired product selectivity in mild reaction condition) were obtained using a sample with weak acid sites.

HMF to 2,5-furandicarboxylic acid as a model reaction for the conversion of renewable molecules was investigated by Bonincontro et al. [4]. Attention was focused on innovative nanofibrous membranes based on Pd-Au catalysts immobilized via electrospinning onto different polymers. The type of polymer and the method used to insert the active phases in the membrane were demonstrated to have a significant effect on catalytic performance. In particular, the authors demonstrated that the hydrophilicity and the glass transition temperature of the polymeric component are key factors for producing active and selective materials. These results underline the promising potential of large-scale applications of electrospinning for the preparation of catalytic nanofibrous membranes to be used in processes for the conversion of renewable molecules.

In their communication, Francy et al. [6] analyzed the performances of nanoscale zero-valent iron (nZVI) particles synthesized from green leaves, which is an example of low-cost biomass. These nanoparticles proved to be effective in the remediation of chlorinated compounds and heavy metals (Pd and Ni) from contaminated soil. Thus, this is an example of the upgrading of biomasses to catalysts, which supports the goal of a circular economy.

Two papers in the Special Issue were devoted to hydrogen production, via partial oxidation of CH_4 [7] or ethanol steam reforming [5].

Regarding methane partial oxidation, Fakeeha et al. investigated Ni, Co, and Ni-Co catalysts supported on binary oxide ZrO_2–Al_2O_3, which were characterized by proper techniques such as XRD, BET, TPR, TPD, TGA, SEM, and TEM. It was observed that increasing the calcination temperature and the addition of ZrO_2 to Al_2O_3 enhances Ni metal-support interaction, improving the catalytic activity and sintering resistance. Furthermore, ZrO_2 provides higher oxygen storage capacity and stronger Lewis basicity, which was observed to contribute to coke suppression, eventually leading to a more stable catalyst.

Regarding ethanol steam reforming, which is one of the most promising ways to produce hydrogen from biomass, a collaboration between groups from Italy and Spain aimed at investigating robust, selective, and active catalysts for this reaction. Pizzolitto et al. studied nickel-ceria catalysts synthesized by two different techniques: microemulsion and precipitation. The effects of lanthanum doping were investigated too. Again, attention was focused on the stability of these samples, because coke deposition is a major issue in these systems.

The stability issue is also investigated in the paper by Nguyen et al. [8], where an iron and nickel bimetallic metal–organic framework material was studied for the Michael addition amidation of 2-aminopyridine and nitroolefins. The catalyst can be reused without a substantial reduction in catalytic activity, with 77% yield after six times of reuse.

Of course, hydrogen is not only a sustainable energy vector, but also a classic green reducing agent. In the paper by Dehghani et al. [9], hydrogen was used in the acetylene hydrogenation process over Pd-Ag supported α-Al_2O_3 catalysts. The work presents an experimental approach to the mechanism-, kinetic-, and decay- related factors in order to predict catalyst activity and proposes a detailed reaction network.

We thank all the contributors for their fundamental support for this Special Issue, as well as the editorial staff of *Processes* for their efforts.

We kindly invite you to read this Special Issue of *Processes*, focused on the following two words: sustainability and catalysis. We believe that through catalysis, it is possible to engineer a greener world.

You can find the papers for free at https://www.mdpi.com/journal/processes/special_issues/metal_nano_catal.

Author Contributions: Both authors have read and agreed to the published version of the Editorial.

Funding: This research received no external funding.

Conflicts of Interest: The authors declare no conflict of interest.

References

1. *Report of the World Commission on Environment and Development (Brundtland Report) Our Common Future*; United Nations: New York, NY, USA, 1987. Available online: https://digitallibrary.un.org/record/139811 (accessed on 5 June 2021).
2. Tran, N.; Uemura, Y.; Trinh, T.; Ramli, A. Hydrodeoxygenation of guaiacol over Pd–Co and Pd–Fe catalysts: Deactivation and regeneration. *Processes* **2021**, *9*, 430. [CrossRef]
3. Taghavi, S.; Ghedini, E.; Menegazzo, F.; Signoretto, M.; Gazzoli, D.; Pietrogiacomi, D.; Matayeva, A.; Fasolini, A.; Vaccari, A.; Basile, F.; et al. MCM-41 Supported Co-Based Bimetallic Catalysts for Aqueous Phase Transformation of Glucose to Biochemicals. *Processes* **2020**, *8*, 843. [CrossRef]
4. Bonincontro, D.; Fraschetti, F.; Squarzoni, C.; Mazzocchetti, L.; Maccaferri, E.; Giorgini, L.; Zucchelli, A.; Gualandi, C.; Focarete, M.L.; Albonetti, S. Pd/Au based catalyst immobilization in polymeric nanofibrous membranes via electrospinning for the selective oxidation of 5-hydroxymethylfurfural. *Processes* **2020**, *8*, 45. [CrossRef]
5. Pizzolitto, C.; Menegazzo, F.; Ghedini, E.; Arias, A.M.; Corberán, V.C.; Signoretto, M. Microemulsion vs. precipitation: Which is the best synthesis of nickel-ceria catalysts for ethanol steam reforming? *Processes* **2020**, *9*, 77. [CrossRef]
6. Francy, N.; Shanthakumar, S.; Chiampo, F.; Sekhar, Y.R. Remediation of lead and nickel contaminated soil using nanoscale zero-valent iron (nZVI) particles synthesized using green leaves: First results. *Processes* **2020**, *8*, 1453. [CrossRef]
7. Fakeeha, A.H.; Arafat, Y.; Ibrahim, A.A.; Shaikh, H.; Atia, H.; Abasaeed, A.E.; Armbruster, U.; Al-Fatesh, A.S. Highly selective syngas/H2 production via partial oxidation of CH4 using (Ni, Co and Ni-Co)/ZrO2-Al2O3 catalysts: Influence of calcination temperature. *Processes* **2019**, *7*, 141. [CrossRef]
8. Nguyen, T.D.; Nguyen, O.K.T.; Van Tran, T.; Nguyen, V.H.; Bach, L.G.; Tran, N.V.; Vo, D.-V.N.; Van Nguyen, T.; Hong, S.-S.; Do, S.T. The synthesis of N-(pyridin-2-yl)-benzamides from aminopyridine and trans-beta-nitrostyrene by Fe2Ni-BDC bimetallic metal–organic frameworks. *Processes* **2019**, *7*, 789. [CrossRef]
9. Dehghani, O.; Rahimpour, M.R.; Shariati, A. An experimental approach on industrial Pd-Ag supported α-Al2O3 catalyst used in acetylene hydrogenation process: Mechanism, kinetic and catalyst decay. *Processes* **2019**, *7*, 136. [CrossRef]

Article

Hydrodeoxygenation of Guaiacol over Pd–Co and Pd–Fe Catalysts: Deactivation and Regeneration

Nga Tran [1,2,*], Yoshimitsu Uemura [1,*], Thanh Trinh [3] and Anita Ramli [1,4]

1. Centre for Biofuel and Biochemical Research, Department of Chemical Engineering, Universiti Teknologi PETRONAS, Seri Iskandar 32610, Perak, Malaysia; anita_ramli@utp.edu.my
2. Key Laboratory of Chemical Engineering and Petroleum Processing, Ho Chi Minh City University of Technology (HCMUT), 268 Ly Thuong Kiet Street, District 10, Ho Chi Minh City 72506, Vietnam
3. Faculty of Food Science and Technology, HCMC University of Food Industry (HUFI), 140 Le Trong Tan, Tan Phu District, Ho Chi Minh City 72009, Vietnam; onalone2000@gmail.com
4. Department of Fundamental & Applied Science, Universiti Teknologi PETRONAS, Seri Iskandar 32610, Perak, Malaysia
* Correspondence: Tonga.tnt@gmail.com (N.T.); YUemura.my@gmail.com (Y.U.); Tel.: +84-985516524 (N.T.)

Abstract: In bio-oil upgrading, the activity and stability of the catalyst are of great importance for the catalytic hydrodeoxygenation (HDO) process. The vapor-phase HDO of guaiacol was investigated to clarify the activity, stability, and regeneration ability of Al-MCM-41 supported Pd, Co, and Fe catalysts in a fixed-bed reactor. The HDO experiment was conducted at 400 °C and 1 atm, while the regeneration of the catalyst was performed with an air flow at 500 °C for 240 min. TGA and XPS techniques were applied to study the coke deposit and metal oxide bond energy of the catalysts before and after HDO reaction. The Co and Pd–Co simultaneously catalyzed the $C_{Ar}O$–CH_3, C_{Ar}–OH, and multiple C–C hydrogenolyses, while the Fe and Pd–Fe principally catalyzed the C_{Ar}–OCH_3 hydrogenolysis. The bimetallic Pd–Co and Pd–Fe showed a higher HDO yield and stability than monometallic Co and Fe, since the coke formation was reduced. The Pd–Fe catalyst presented a higher stability and regeneration ability than the Pd–Co catalyst, with consistent activity during three HDO cycles.

Keywords: hydrodeoxygenation; guaiacol; regeneration; catalyst deactivation

1. Introduction

The lignocellulose biomass resource can be used not only as direct energy in combustion, but also as a more valuable fuel after the conversion and upgrading process [1]. Pyrolysis is a thermal conversion of biomass to produce bio-oil, which has significant advantages in storage, transportation, and the ability to be utilized as useful petrochemicals and fuel [2]. However, the presence of oxygenated compounds (e.g., acids, esters, alcohols, ketones, furans, and phenols) gives the bio-oil a low heating value, low chemical and thermal stability, high viscosity, and high corrosiveness [3–7]. These disadvantages can be mitigated or solved if oxygen is removed partially or entirely, respectively [8]. Catalytic hydrodeoxygenation (HDO) is a prominent process for bio-oil upgrading, since it can eliminate the oxygen significantly and preserve the carbon of the bio-oil [9,10].

The stability and regeneration abilities of catalysts are very important in the catalytic HDO process. In the HDO process, the deactivation of catalysts is mainly from coke deposits, sintering, poisoning, and metal deposition [8,11,12]. Coke deposits are formed through polymerization and polycondensation reactions on the catalytic surface, resulting in pore blockages and active site coverage [8]. Water and S- or N-containing compounds in the feed can cause poisoning on the catalytic surface [13]. Sintering is the agglomeration of nanoparticles into larger particles, resulting in a decrease in the active sites [14]. In the hydrotreating of different bio-oil sources over different catalyst types (e.g., guaiacol over noble metal catalysts [15], grass bio-oil over noble metal Ru and Pt [16], rice husk bio-oil over Ni–Cu catalyst [17], or pine bio-oil over NiAl$_2$O$_4$ [18]), the coke deposit is the main cause of the catalyst deactivation. The coke deposit is dependent on the catalyst type, feedstock, and operating conditions [17]. The deactivated catalysts can be regenerated via coke combustion at medium to high temperatures, depending on the HDO reaction conditions [16]. In a catalyst HDO, the mesoporous supports exhibited much higher stability than the microporous supports [18,19]. There are numerous research studies on catalyst deactivation effects, e.g., the type of carbon deposit, metal sintering, deactivation mechanism, and bio-oil impurities (H$_2$O, H$_2$S, etc.) [14,16,20–22]. However, the regeneration abilities of catalysts during catalytic HDO are not well understood and have only been examined in a few studies [9,23,24].

In this study, the HDO of guaiacol on Al-MCM-41 supported Pd–Co and Pd–Fe catalysts were investigated in a fixed-bed, continuous-flow reactor at ambient pressure. The Al-MCM-41 is an acidic and mesoporous support, which can enhance the transalkylation activity and stability of the catalyst in the HDO process [18,25,26]. Guaiacol was chosen as a model compound because it contains both major functional groups of lignin-derived phenolic, such as hydroxyl (–OH) and methoxy (–OCH$_3$) groups. The HDO of guaiacol was conducted to screen the HDO activity, stability, and regeneration ability of the catalysts. TGA and XPS were applied to characterize the deactivation that occurred during catalytic HDO.

2. Materials and Methods

2.1. Materials

Mesoporous aluminosilicate Al-MCM-41 support (3–4% Al$_2$O$_3$) was supplied by ACS Material (Pasadena, CA, US). Guaiacol (2-methoxyphenol) purchased from Merk (Kenilworth, NJ, US) was used as the model compound for the HDO study. Metal precursors (palladium(II) nitrate, cobalt(II) nitrate (99.999%), and iron(III) nitrate (99.95%)) were purchased from Aldrich (St. Louis, MO, US).

2.2. Catalyst Preparation and Characterization

The catalysts were prepared via an incipient wetness co-impregnation method. The detailed characterization of the catalyst has been previously described [27]. The Al-MCM-41 supported catalysts had a mesoporosity structure, with a pore size of around 3 nm. The total acidity of the Al-MCM-41 support measured by temperature programmed desorption (TPD) of ammonia was 1.06 mmol/g. The transmission electron microscopy (TEM), temperature programmed reduction (TPR) in hydrogen, and powder X-ray diffraction (XRD) results implied that the addition of Pd could improve the dispersion and reducibility of Co and Fe oxides with the formation of Pd–Co and Pd–Fe alloys.

Thermogravimetric analysis (TGA) under the flow of air was conducted in a TA Instrument model QA50. During the TGA analysis, temperature was increased from room temperature to 900 °C, at a heating rate of 10 °C/min. X-ray photoelectron spectroscopy (XPS) was performed with a Thermo Scientific K-Alpha system equipped with an Al Kα radiation source. The spectrometer was operated with the constant analyzer energy (CAE) mode at a pass energy of 50 eV and a step of 0.1 eV. Quantification and deconvolution were performed using the Gaussian functions of the OriginPro 2015 software (OriginLab, Northampton, MA, US).

2.3. HDO of Guaiacol

A Catalytic HDO reaction was conducted in a fixed-bed reactor at 400 °C and ambient pressure. The details of the experimental set-up of the HDO of guaiacol were mentioned in a previous report [27]. Before the HDO reaction, all the catalysts were reduced to 450 °C using a hydrogen flow of 90 mL/min for 2 h. Pure guaiacol was fed at a flow rate of 1.08 mL/h using a syringe pump and vaporized at 350 °C in the top glass wool bed. Catalyst regeneration was carried out after 210 min of guaiacol HDO reaction. The used catalyst was first treated with an air flow at 500 °C for 240 min. Afterwards, the catalyst was reactivated in hydrogen flow at 450 °C for 120 min and catalyzed a new HDO reaction cycle. The liquid products were quantified by a Shimadzu GC-2014 gas chromatography (GC), with a SGE BPX–5 capillary column (30 m, ID 0.25 mm, and 0.25 µm) and a flame ionized detector (FID). The gas products were analyzed by a Shimadzu GC–8A system equipped with a thermal conductivity detector (TCD). The carbon balance of the HDO experiments was between 93% and 98%. The HDO of guaiacol over the Pd–Fe catalyst at W/F of 1.67 h and temperature of 400 °C were repeated twice, and the standard deviation of all product yields was less than 1.0 Mol$_C$%. Meanwhile, the other HDO experiments were conducted once. Carbon-based guaiacol conversion (X_{Gua}), product yields (Y_i), and HDO yields were calculated in Mol$_{Carbon}$% by the following equations.

$$X_{Gua}\,(\%) = \frac{Mol(gua)_{in} - Mol(gua)_{out}}{Mol(gua)_{in}} \times 100 \quad (1)$$

$$Y_i\,(\%) = \frac{Mol_i \times \alpha_i}{Mol(gua)_{in} \times \alpha_{gua}} \times 100 \quad (2)$$

$$HDO\ yield\,(\%) = \sum_{i=1}^{25} \frac{Y_i \times (\beta_{gua} - \beta_i)}{\beta_{gua}} \quad (3)$$

where α_i and β_i are the carbon and oxygen numbers in the product i; α_{gua} = 7 and β_{gua} = 2.

3. Results

3.1. Catalytic Stability of Mono- and Bimetallic

Figure 1 compares the conversion of guaiacol and product yields over the supported mono- and bimetallic catalysts with time on stream (TOS). The monometallic Fe catalyst had higher mono-oxygenated products and lower gas phase (which mainly contained methane) yields than monometallic Co, resulting in higher HDO yield. Addition of Pd to the Co catalyst increased the guaiacol conversion and deoxygenated product (aromatics and mono-oxygenated) yields. However, this addition to the Fe catalyst only showed the increment of guaiacol conversion and mono-oxygenated product yield, while the oxygen-free aromatic yield was decreased. These implied that the Fe active sites mainly catalyzed the C_{Ar}–OCH_3 cleavage reaction instead of the C_{Ar}–OH cleavage and produced mono-oxygenate as the main product. Meanwhile, the hydrogenolysis of C_{Ar}–OR and C–C groups occurred simultaneously in the HDO over the Pd–Co and Co catalysts, resulting in the formation of deoxygenated products and methane. The reaction routes of HDO of guaiacol over different catalysts can be found in Scheme 1. As shown in Figure 1, the monometallic Co and Fe catalysts showed a faster deactivation than the bimetallic Pd–Co and Pd–Fe. The addition of Pd significantly enhanced the stability of both Co and Fe catalysts. Among these catalysts, Pd–Fe presented as the most promising catalyst due to its higher stability and HDO yield. In summary, the addition of Pd enhanced the guaiacol conversion, HDO yield, and stability of the Co and Fe catalysts. Previous studies mentioned the enhancement in conversion and HDO yields when novel metals (Pd and Pt) were added [9,28,29]. However, there was no report on the stability enhancement like in our findings (Supplementary materials).

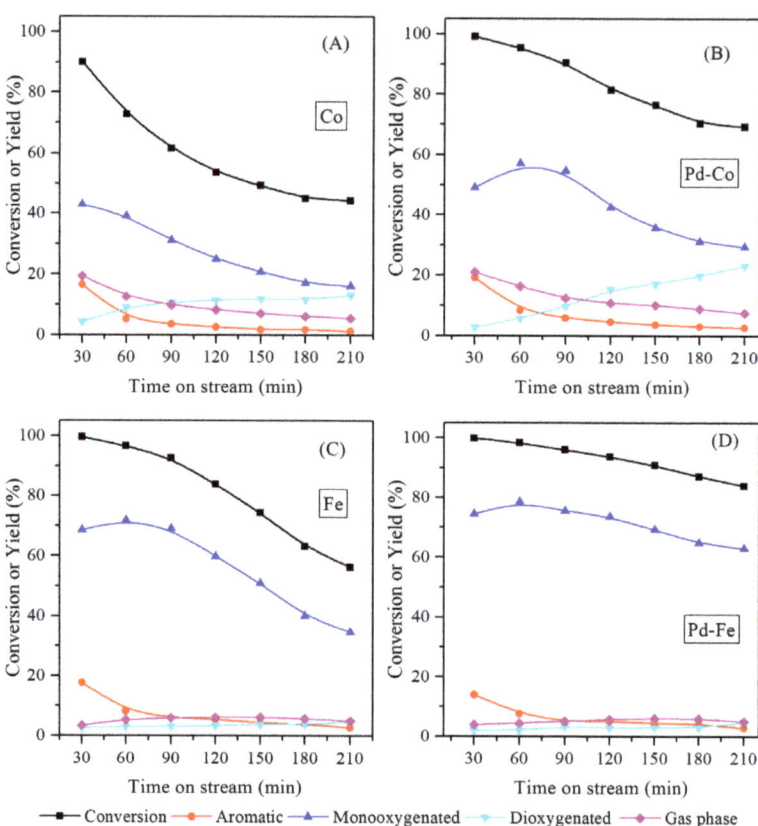

Figure 1. Catalytic hydrodeoxygenation (HDO) of guaiacol over mono- and bimetallic catalysts (**A**) Co, (**B**) Pd-Co, (**C**) Fe and (**D**) Pd-Fe. Reaction conditions: T = 400 °C, P = 1 bar, H_2/Gua = 25, W/F = 0.83 h.

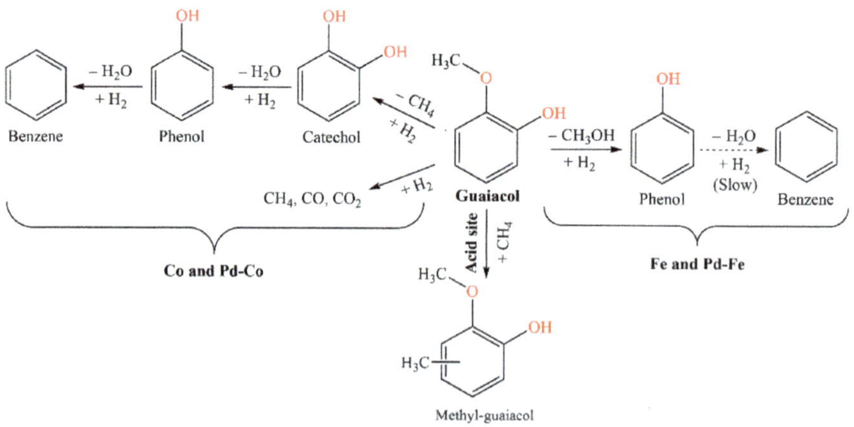

Scheme 1. Simple reaction pathways of HDO of guaiacol. More details can be found elsewhere [27].

To understand the contribution of Pd to the stability of Co and Fe catalysts, the TGA of the used catalysts was applied to understand the coke formation in catalysts, as shown in Figure 2. There was a negative peak, which appeared at around 200 °C in the derivative thermogravimetric (DTG) curves (in Figure 2A). This peak could be attributed to the oxidation of the remaining metallic Fe or Co, which were reduced during the HDO reaction. The main peaks in the DTG curves were observed from 200 to 650 °C, which was associated with coke removal by oxidation. These mass loss data of used mono- and bimetallic catalysts are compared in Figure 2B. The bimetallic Pd–Co and Pd–Fe catalysts had a lower coke formation than the corresponding monometallic catalysts. In summary, the addition of Pd prevented coke formation during HDO reactions and made the catalyst more stable. As shown in Figure 2B, the used Fe and Pd–Fe catalysts had a higher coke formation than the used Co and Pd–Co catalysts. Nevertheless, the stability of Fe and Pd–Fe catalysts was higher than that of Co and Pd–Co catalysts (Figure 1). This contradiction could be explained by the DTG results, in which the used Fe and Pd–Fe catalysts had lower temperature degradation peaks (i.e., 350 °C) than Co and Pd–Co ones (i.e., 500 °C). Hence, the coke formation during HDO over Fe and Pd–Fe was more easily degraded than the one over Co and Pd–Co.

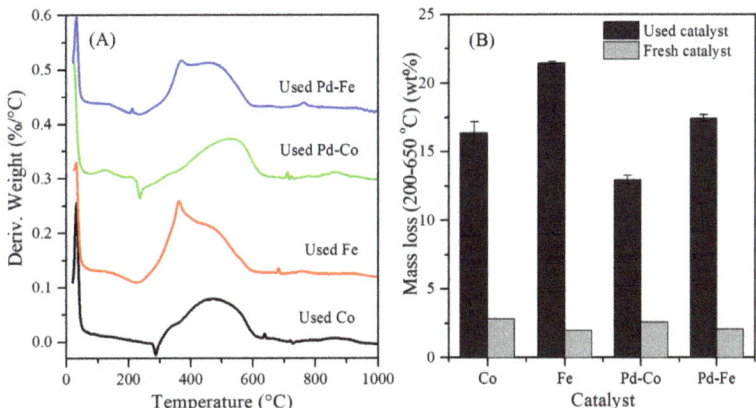

Figure 2. TG results of mono- and bimetallic catalysts. (**A**) DTG curves of used catalysts in air atmosphere, (**B**) Mass loss of used and fresh catalysts.

3.2. Regeneration of Bimetallic Catalysts

During HDO reactions, catalysts are deactivated due to carbon deposition, sintering, or poisoning; hence, regeneration ability becomes an important issue in practical applications [14,22]. In our previous study [30], with the regeneration of air at 450 °C for 2 h, the coke deposit remained on the catalyst surface. Hence, the treatment temperature was increased to 500 °C, and the time was prolonged to 4 h in order to improve the regenerated catalyst in this current work. Figure 3 illustrates the details of the regeneration ability of the Pd–Co and Pd–Fe catalysts. During the first HDO reaction, Pd–Fe and Pd–Co had the same guaiacol conversion. However, Pd–Fe presented a higher HDO yield than Pd–Co, due to its lower gasification activity. The regenerated Pd–Co catalyst showed a decrease in guaiacol conversion and HDO yield compared with the fresh one. In addition, this regenerated catalyst gave a faster deactivation than the fresh one, and the deactivation rate increased with the increase in recycle time. In contrast to Pd–Co, Pd–Fe had considerably higher stability and regeneration ability. The regenerated Pd–Fe catalyst had a higher HDO yield than the fresh one. Even at the 3rd cycle, the Pd–Fe catalyst gave no significant deactivation after a 210 min reaction. In summary, the Pd–Fe catalyst could be regenerated due to there being no significant change in the guaiacol conversion and HDO yield during the three reaction cycles.

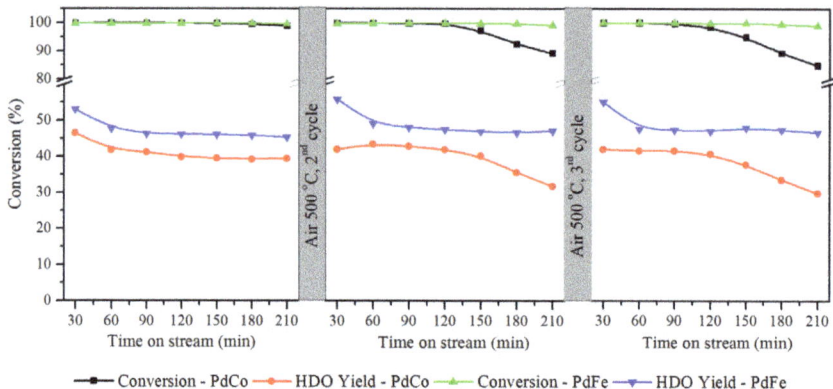

Figure 3. Recycling of HDO of guaiacol over Pd–Co and Pd–Fe catalysts (Reduction–Reaction–Regeneration). Reaction condition: T = 400 °C, P = 1 bar, H_2/Gua = 25, W/F = 1.67 h.

The regenerations of catalysts in previous studies were conducted in air at 350 °C [23] or 500 °C [24,26]. There were slight changes in the conversion and product selectivity of the used catalyst. When the catalyst was treated in air at 350 °C, the coke deposits on the Pt catalysts remained during the regeneration, which resulted in a slight decrease in m-cresol conversion and higher selectivity to toluene in the second cycle [23]. In our current report, the treatment with air at 500 °C could even increase the HDO yield of Pd–Fe catalyst.

The addition of Pd can improve the stability of the Fe and Co catalyst. Moreover, the Pd–Fe catalyst showed considerably higher stability and regeneration ability than the Pd–Co catalyst. TGA and XPS analysis were applied to study the catalyst deactivation and regeneration during HDO reactions. After three cycles of HDO reaction, the used Pd–Co and Pd–Co catalysts were taken out and regenerated with air at 500 °C for 4 h in the muffle furnace. These fresh, used, and regenerated catalysts were analyzed with XPS and TGA to clarify the deactivation of the catalyst. According to the DTG results in Figure 2A, the coke deposits on Pd–Fe degraded at a lower temperature than Pd–Co, resulting in a higher regeneration ability of Pd–Fe catalysts.

Using the XPS spectra of Al-MCM-41 support, fresh (calcined), reduced, used, and regenerated catalysts were plotted and compared to reveal the change of elemental components and chemical state of the catalyst. Figures 4 and 5 illustrate the deconvoluted XPS spectra of Pd–Co and Pd–Fe catalysts, respectively. The XPS spectra of other elements (Si 2p, Al 2p, and Pd 3d) can be found in Figures S1 and S2. All spectra were calibrated by referring to the maximum of the O 1s peak at 533.0 eV, which corresponded to Si–O–Si binding in SiO_4 species [31–34].

Figures 4A and 5A show the C 1s XPS spectra of reduced, used, and regenerated Pd–Co and Pd–Fe, respectively. The used catalyst surface was covered with a carbon deposit, which formed during HDO reaction; hence, the carbon signal of the used catalyst was higher than the other catalysts, while the metal signals (Co 2p and Fe 2p) of the used catalyst were lower than others. C 1s spectra of used Pd–Co had two distinct peaks at 284.7 and 282.1 eV, whereas the used Pd–Fe had one additional peak at 280.0 eV. The peak at 284.7 eV was attributed to contaminated carbon, which appeared on all reduced, used, and regenerated catalysts [35,36]. The peak at 282.3 eV in used catalysts could be assigned to graphite-like carbon [35–39]. According to previous papers on coke deposits in used catalysts [35,37–39], the peaks of oxidized carbon should appear at a higher binding energy position than the contaminated carbon peak. These oxidized carbon peaks were absent in our used Pd–Co and Pd–Fe catalysts. The additional peaks at 280.0 eV in the used Pd–Fe catalyst could be attributed to dehydrogenated carbon species [38,39]. In general, the regeneration process can remove the carbon deposit on the catalyst surface significantly.

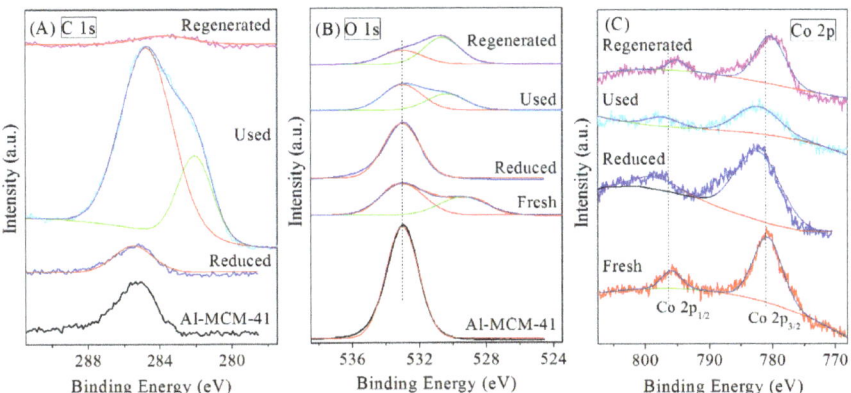

Figure 4. The deconvoluted core level XPS scans of C1s (**A**), O1s (**B**), and Co2p (**C**) of fresh, reduced, used, and regenerated Pd–Co/Al-MCM-41 catalysts.

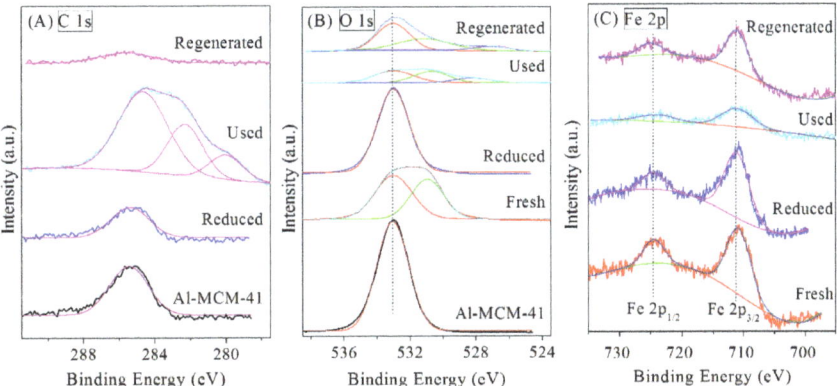

Figure 5. The deconvoluted core level XPS scans of C1s (**A**), O1s (**B**), and Fe2p (**C**) of fresh, reduced, used, and regenerated Pd–Fe/Al-MCM-41 catalysts.

In Figures 4B and 5B, the O 1s spectra of support and reduced catalysts had only one peak at 533.0 eV, while other catalysts had additional peaks at lower binding energy. The profile of the O 1s spectra (Figures 4B and 5B) was extremely similar to the profile of Si 2p and Al 2p spectra (see Figures S1 and S2) for all catalysts; this implies that the oxygen would bond with at least one silicon or aluminum atom (small amount) [33,34]. The first peak at 533.0 eV might correspond to Si–O–Si binding in SiO_4 species [31–34], while the second peak at 531 to 529 eV could be ascribed to the Si–O–Me bindings (Me = Co or Fe). The second peak in fresh catalyst corresponds to Si–O–Me bindings, since it disappeared after the catalysts were reduced. The second peak of fresh Pd–Co appeared at 529 eV, whereas the fresh Pd–Fe appeared at a higher energy binding of 531 eV. The second peak of the Pd–Co catalyst was shifted to higher binding energy after the catalyst was used and regenerated; this might explain the drop in HDO activity of this catalyst. The O 1s oxidation state of Pd–Co did not change as much as Pd–Fe between fresh, used, and regenerated catalysts. The fresh Pd–Fe had two peaks at 533 and 531 eV; however, the used and regenerated Pd–Fe formed a new peak at a lower binding energy (528 or 527 eV). This formation of a lower binding energy peak of Pd–Fe might be related to the enhancement in HDO activity in the regenerated catalyst, as discussed above.

As can be seen in Figure 4C, the binding energies of Co $2p_{3/2}$ and Co $2p_{1/2}$ appeared as two distinct peaks, at around 781.0 and 796.5 eV. The spin–orbit splitting energy ($\Delta E_{Co2p1/2-Co2p3/2}$) of the fresh, used, and regenerated catalysts was about 15.1 eV, combining with the very weak satellites. This observation indicated the coexistence of Co^{2+} and Co^{3+} in this Pd–Co catalyst [40,41]. The reduced catalyst had higher spin–orbit splitting energy (15.5 eV) and shake-up satellite than others. The surface of the used catalyst was covered by a carbon deposit; hence, the signal of the metal active site was smaller than that of the fresh, reduced, and regenerated catalysts. The regeneration process can remove the carbon deposit; however, the intensity of Co 2p peaks was not as high as that of fresh and reduced catalysts.

The XPS spectra of Fe 2p (Figure 5C) had two main peaks, which can be assigned as Fe $2p_{3/2}$ at 711.2 eV and Fe $2p_{1/2}$ at 725.0 eV [31]. The maximum Fe $2p_{3/2}$ peak was observed at around 711.0 eV with the satellites at higher binding energy, suggesting that Fe species were mainly in Fe^{3+} state [31,42,43]. Similar to the Pd–Co catalyst, the used and regenerated Pd–Fe catalyst had a lower signal intensity of Fe 2p compared to a fresh and reduced catalyst, due to the effect of coke deposit. However, the peak position of the Pd–Fe catalyst was not changed significantly, like the Pd–Co catalyst.

In summary, the XPS result reveals that the air treatment at 500 °C for 4 h could remove most of the coke deposit on the catalyst. Moreover, the Si–O–Co binding of used and regenerated Pd–Co catalyst shifted to higher binding energy, resulting in the drop of HDO activity in the second and third cycles. In addition, the used and regenerated Pd–Fe formed a new peak at lower binding energy, yielding the enhancement in HDO activity of the regenerated catalyst.

4. Conclusions

Hydrodeoxygenation of guaiacol over Al-MCM-41 supported Pd–Co and Pd–Fe catalysts were studied at 400 °C and ambient atmosphere. The Fe catalyst gave a higher HDO yield and lower gas-phase yield compared with the Co catalyst in HDO of guaiacol. The bimetallic Pd–Co and Pd–Fe achieved a higher conversion and HDO yield than the monometallic Co and Fe. Interestingly, the addition of Pd significantly improved the stability of the catalysts, since it could suppress the coke deposition on the catalysts. Furthermore, the Pd–Fe catalyst presented a higher stability and regeneration ability than the Pd–Co catalyst. The coke deposits were mostly removed by the treatment at 500 °C in air, which was confirmed by TGA and XPS results. The regenerated Pd–Co catalyst showed a decrease in HDO yield and stability, while the Pd–Fe catalyst presented consistent activity during three HDO cycles. This can be explained by the lower thermal stability coke deposit and the formation of lower binding energy Si–O–Fe bonds of the used Pd–Fe catalyst.

Supplementary Materials: The following are available online at https://www.mdpi.com/2227-9717/9/3/430/s1, Figure S1: De-convoluted Si 2p, Al 2p and Pd 3d XPS spectra of fresh, used, and regenerated Pd–Co/Al-MCM-41catalysts; Figure S2: De-convoluted Si 2p, Al 2p and Pd 3d XPS spectra of fresh, used, and regenerated Pd–Fe/Al-MCM-41catalysts.

Author Contributions: Conceptualization, N.T. and Y.U.; methodology, N.T.; software, T.T.; formal analysis, N.T.; investigation, N.T.; writing—original draft preparation, N.T.; writing—review and editing, Y.U., A.R., and T.T.; supervision, Y.U. and A.R.; project administration, Y.U.; funding acquisition, Y.U. All authors have read and agreed to the published version of the manuscript.

Funding: This research was funded by Mitsubishi Corporation Educational Trust Fund.

Data Availability Statement: Data sharing is not applicable.

Acknowledgments: We wish to thank Universiti Teknologi PETRONAS for providing a congenial work environment and state-of-the-art research facilities. We also thank Masaharu Komiyama for his valuable support, comments, and suggestions. We acknowledge the support, both in terms of time and financially, from Ho Chi Minh City University of Technology (HCMUT).

Conflicts of Interest: The authors declare no conflict of interest.

References

1. Bridgwater, A. *Thermal Biomass Conversion and Utilization: Biomass Information System*; EUR; 16863 EN.; Office for Official Publications of the European Communities: Luxembourg, 1996; ISBN 92-827-7207-1.
2. Torri, I.D.V.; Paasikallio, V.; Faccini, C.S.; Huff, R.; Caramão, E.B.; Sacon, V.; Oasmaa, A.; Zini, C.A. Bio-Oil Production of Softwood and Hardwood Forest Industry Residues through Fast and Intermediate Pyrolysis and Its Chromatographic Characterization. *Bioresour. Technol.* **2016**, *200*, 680–690. [CrossRef]
3. Koike, N.; Hosokai, S.; Takagaki, A.; Nishimura, S.; Kikuchi, R.; Ebitani, K.; Suzuki, Y.; Oyama, S.T. Upgrading of Pyrolysis Bio-Oil Using Nickel Phosphide Catalysts. *J. Catal.* **2016**, *333*, 115–126. [CrossRef]
4. Olarte, M.V.; Zacher, A.H.; Padmaperuma, A.B.; Burton, S.D.; Job, H.M.; Lemmon, T.L.; Swita, M.S.; Rotness, L.J.; Neuenschwander, G.N.; Frye, J.G.; et al. Stabilization of Softwood-Derived Pyrolysis Oils for Continuous Bio-Oil Hydroprocessing. *Top. Catal.* **2015**, 1–10. [CrossRef]
5. Lehto, J.; Oasmaa, A.; Solantausta, Y.; Kytö, M.; Chiaramonti, D. Review of Fuel Oil Quality and Combustion of Fast Pyrolysis Bio-Oils from Lignocellulosic Biomass. *Appl. Energy* **2014**, *116*, 178–190. Available online: https://cris.vtt.fi/en/publications/fuel-oil-quality-and-combustion-of-fast-pyrolysis-bio-oils (accessed on 26 February 2021). [CrossRef]
6. Uemura, Y.; Omar, W.N.; Razlan, S.; Afif, H.; Yusup, S.; Onoe, K. Mass and Energy Yields of Bio-Oil Obtained by Microwave-Induced Pyrolysis of Oil Palm Kernel Shell. *J. Jpn. Inst. Energy* **2012**, *91*, 954–959. [CrossRef]
7. Meng, J.; Moore, A.; Tilotta, D.C.; Kelley, S.S.; Adhikari, S.; Park, S. Thermal and Storage Stability of Bio-Oil from Pyrolysis of Torrefied Wood. *Energy Fuels* **2015**, *29*, 5117–5126. [CrossRef]
8. Mortensen, P.M.; Grunwaldt, J.-D.; Jensen, P.A.; Knudsen, K.G.; Jensen, A.D. A Review of Catalytic Upgrading of Bio-Oil to Engine Fuels. *Appl. Catal. A Gen.* **2011**, *407*, 1–19. [CrossRef]
9. Sun, J.; Karim, A.M.; Zhang, H.; Kovarik, L.; Li, X.S.; Hensley, A.J.; McEwen, J.-S.; Wang, Y. Carbon-Supported Bimetallic Pd–Fe Catalysts for Vapor-Phase Hydrodeoxygenation of Guaiacol. *J. Catal.* **2013**, *306*, 47–57. [CrossRef]
10. Nie, L.; de Souza, P.M.; Noronha, F.B.; An, W.; Sooknoi, T.; Resasco, D.E. Selective Conversion of m-Cresol to Toluene over Bimetallic Ni–Fe Catalysts. *J. Mol. Catal. A Chem.* **2014**, *388–389*, 47–55. [CrossRef]
11. Bartholomew, C.H.; Farrauto, R.J. *Fundamentals of Industrial Catalytic Processes*; John Wiley & Son: Hoboken, NJ, USA, 2005; ISBN 978-0-471-45713-8.
12. Saidi, M.; Samimi, F.; Karimipourfard, D.; Nimmanwudipong, T.; Gates, B.C.; Rahimpour, M.R. Upgrading of Lignin-Derived Bio-Oils by Catalytic Hydrodeoxygenation. *Energy Environ. Sci.* **2013**, *7*, 103–129. [CrossRef]
13. Bridgwater, A.V. Review of Fast Pyrolysis of Biomass and Product Upgrading. *Biomass-Bioenergy* **2012**, *38*, 68–94. [CrossRef]
14. Contreras, J.L.; Fuentes, G.A. *Sintering of Supported Metal Catalysts, Sintering-Methods and Products, Volodymyr Shatokha*; IntechOpen: London, UK, 2012. [CrossRef]
15. Gao, D.; Schweitzer, C.; Hwang, H.T.; Varma, A. Conversion of Guaiacol on Noble Metal Catalysts: Reaction Performance and Deactivation Studies. *Ind. Eng. Chem. Res.* **2014**, *53*, 18658–18667. [CrossRef]
16. Oh, S.; Hwang, H.; Choi, H.S.; Choi, J.W. The Effects of Noble Metal Catalysts on the Bio-Oil Quality during the Hydrodeoxygenative Upgrading Process. *Fuel* **2015**, *153*, 535–543. [CrossRef]
17. Li, Y.; Zhang, C.; Liu, Y.; Tang, S.; Chen, G.; Zhang, R.; Tang, X. Coke Formation on the Surface of Ni/HZSM-5 and Ni-Cu/HZSM-5 Catalysts during Bio-Oil Hydrodeoxygenation. *Fuel* **2017**, *189*, 23–31. [CrossRef]
18. Remiro, A.; Valle, B.; Aguayo, A.T.; Bilbao, J.; Gayubo, A.G. Operating Conditions for Attenuating Ni/La$_2$O$_3$ –AAl$_2$O$_3$ Catalyst Deactivation in the Steam Reforming of Bio-Oil Aqueous Fraction. *Fuel Process. Technol.* **2013**, *115*, 222–232. [CrossRef]
19. Wang, Y.; Fang, Y.; He, T.; Hu, H.; Wu, J. Hydrodeoxygenation of Dibenzofuran over Noble Metal Supported on Mesoporous Zeolite. *Catal. Commun.* **2011**, *12*, 1201–1205. [CrossRef]
20. Li, W.; Li, F.; Wang, H.; Liao, M.; Li, P.; Zheng, J.; Tu, C.; Li, R. Hierarchical Mesoporous ZSM-5 Supported Nickel Catalyst for the Catalytic Hydrodeoxygenation of Anisole to Cyclohexane. *Mol. Catal.* **2020**, *480*, 110642. [CrossRef]
21. Mortensen, P.M.; Gardini, D.; Damsgaard, C.D.; Grunwaldt, J.-D.; Jensen, P.A.; Wagner, J.B.; Jensen, A.D. Deactivation of Ni-MoS2 by Bio-Oil Impurities during Hydrodeoxygenation of Phenol and Octanol. *Appl. Catal. A Gen.* **2016**, *523*, 159–170. [CrossRef]
22. Xu, X.; Jiang, E.; Du, Y.; Li, B. BTX from the Gas-Phase Hydrodeoxygenation and Transmethylation of Guaiacol at Room Pressure. *Renew. Energy* **2016**, *96*(Part A), 458–468. [CrossRef]
23. Olcese, R.; Bettahar, M.M.; Malaman, B.; Ghanbaja, J.; Tibavizco, L.; Petitjean, D.; Dufour, A. Gas-Phase Hydrodeoxygenation of Guaiacol over Iron-Based Catalysts. Effect of Gases Composition, Iron Load and Supports (Silica and Activated Carbon). *Appl. Catal. B Environ.* **2013**, *129*, 528–538. [CrossRef]
24. Zanuttini, M.S.; Costa, B.O.D.; Querini, C.A.; Peralta, M.A. Hydrodeoxygenation of m-Cresol with Pt Supported over Mild Acid Materials. *Appl. Catal. A Gen.* **2014**, *482*, 352–361. [CrossRef]
25. Bi, P.; Yuan, T.; Fan, M.; Jiang, P.; Zhai, Q.; Li, Q. Production of Aromatics through Current-Enhanced Catalytic Conversion of Bio-Oil Tar. *Bioresour. Technol.* **2013**, *136*, 222–229. [CrossRef] [PubMed]
26. Zhu, X.; Lobban, L.L.; Mallinson, R.G.; Resasco, D.E. Bifunctional Transalkylation and Hydrodeoxygenation of Anisole over a Pt/HBeta Catalyst. *J. Catal.* **2011**, *281*, 21–29. [CrossRef]
27. Selvaraj, M.; Shanthi, K.; Maheswari, R.; Ramanathan, A. Hydrodeoxygenation of Guaiacol over MoO$_3$-NiO/Mesoporous Silicates: Effect of Incorporated Heteroatom. *Energy Fuels* **2014**, *28*, 2598–2607. [CrossRef]

28. Tran, N.T.T.; Uemura, Y.; Ramli, A.; Trinh, T.H. Vapor-Phase Hydrodeoxygenation of Lignin-Derived Bio-Oil over Al-MCM-41 Supported Pd-Co and Pd-Fe Catalysts. *Mol. Catal.* **2021**, 111435. [CrossRef]
29. Do, P.T.M.; Foster, A.J.; Chen, J.; Lobo, R.F. Bimetallic Effects in the Hydrodeoxygenation of Meta-Cresol on γ-Al_2O_3 Supported Pt–Ni and Pt–Co Catalysts. *Green Chem.* **2012**, *14*, 1388–1397. [CrossRef]
30. Lai, Q.; Zhang, C.; Holles, J.H. Hydrodeoxygenation of Guaiacol over Ni@Pd and Ni@Pt Bimetallic Overlayer Catalysts. *Appl. Catal. A Gen.* **2016**, *528*, 1–13. [CrossRef]
31. Tran, N.T.T.; Uemura, Y.; Chowdhury, S.; Ramli, A. Vapor-Phase Hydrodeoxygenation of Guaiacol on Al-MCM-41 Supported Ni and Co Catalysts. *Appl. Catal. A Gen.* **2016**, *512*, 93–100. [CrossRef]
32. Yan, Y.; Jiang, S.; Zhang, H.; Zhang, X. Preparation of Novel Fe-ZSM-5 Zeolite Membrane Catalysts for Catalytic Wet Peroxide Oxidation of Phenol in a Membrane Reactor. *Chem. Eng. J.* **2015**, *259*, 243–251. [CrossRef]
33. Bukallah, S.B.; Bumajdad, A.; Khalil, K.M.S.; Zaki, M.I. Characterization of Mesoporous VO_x/MCM-41 Composite Materials Obtained via Post-Synthesis Impregnation. *Appl. Surf. Sci.* **2010**, *256*, 6179–6185. [CrossRef]
34. Méndez, F.J.; Franco-López, O.E.; Bokhimi, X.; Solís-Casados, D.A.; Escobar-Alarcón, L.; Klimova, T.E. Dibenzothiophene Hydrodesulfurization with NiMo and CoMo Catalysts Supported on Niobium-Modified MCM-41. *Appl. Catal. B Environ.* **2017**, *219*, 479–491. [CrossRef]
35. Misran, H.; Salim, M.A.; Ramesh, S. Effect of Ag Nanoparticles Seeding on the Properties of Silica Spheres. *Ceram. Int.* **2018**, *44*, 5901–5908. [CrossRef]
36. Jiang, L.; Li, H.; Wang, Y.; Ma, W.; Zhong, Q. Characterization of Carbon Deposits on Coked Lithium Phosphate Catalysts for the Rearrangement of Propylene Oxide. *Catal. Commun.* **2015**, *64*, 22–26. [CrossRef]
37. Tian, Y.-P.; Liu, X.-M.; Rood, M.J.; Yan, Z.-F. Study of Coke Deposited on a VO_x-K_2O/γ-Al_2O_3 Catalyst in the Non-Oxidative Dehydrogenation of Isobutane. *Appl. Catal. A Gen.* **2017**, *545*, 1–9. [CrossRef]
38. Mao, A.; Wang, H.; Pan, R. Coke Deactivation of Activated Carbon-Supported Rubidium–Potassium Catalyst for C_2F_5I Gas-Phase Synthesis. *J. Fluor. Chem.* **2013**, *150*, 21–24. [CrossRef]
39. Bai, T.; Zhang, X.; Wang, F.; Qu, W.; Liu, X.; Duan, C. Coking Behaviors and Kinetics on HZSM-5/SAPO-34 Catalysts for Conversion of Ethanol to Propylene. *J. Energy Chem.* **2016**, *25*, 545–552. [CrossRef]
40. Gou, M.-L.; Cai, J.; Song, W.; Liu, Z.; Ren, Y.-L.; Pan, B.; Niu, Q. Coking and Deactivation Behavior of ZSM-5 during the Isomerization of Styrene Oxide to Phenylacetaldehyde. *Catal. Commun.* **2017**, *98*, 116–120. [CrossRef]
41. Younis, A.; Chu, D.; Lin, X.; Lee, J.; Li, S. Bipolar Resistive Switching in P-Type Co3O4 Nanosheets Prepared by Electrochemical Deposition. *Nanoscale Res. Lett.* **2013**, *8*, 36. [CrossRef] [PubMed]
42. Lin, Q.; Zhang, Q.; Yang, G.; Chen, Q.; Li, J.; Wei, Q.; Tan, Y.; Wan, H.; Tsubaki, N. Insights into the Promotional Roles of Palladium in Structure and Performance of Cobalt-Based Zeolite Capsule Catalyst for Direct Synthesis of C5–C11 Iso-Paraffins from Syngas. *J. Catal.* **2016**, *344*, 378–388. [CrossRef]
43. Wang, L.; Pu, C.; Xu, L.; Cai, Y.; Guo, Y.; Guo, Y.; Lu, G. Effect of Supports over Pd/Fe_2O_3 on CO Oxidation at Low Temperature. *Fuel Process. Technol.* **2017**, *160*, 152–157. [CrossRef]

Article

MCM-41 Supported Co-Based Bimetallic Catalysts for Aqueous Phase Transformation of Glucose to Biochemicals

Somayeh Taghavi [1], Elena Ghedini [1], Federica Menegazzo [1], Michela Signoretto [1,*], Delia Gazzoli [2], Daniela Pietrogiacomi [2], Aisha Matayeva [3], Andrea Fasolini [3], Angelo Vaccari [3], Francesco Basile [3] and Giuseppe Fornasari [3]

1. CATMAT Lab, Department of Molecular Sciences and Nanosystems, Ca' Foscari University, Venice and INSTM-RUVe, 155 Via Torino, 30172 Venezia Mestre, Italy; somayeh.taghavi@unive.it (S.T.); gelena@unive.it (E.G.); fmenegaz@unive.it (F.M.)
2. Department of Chemistry, Sapienza University of Rome, 5 P.le A.Moro, 00185 Rome, Italy; delia.gazzoli@uniroma1.it (D.G.); daniela.pietrogiacomi@uniroma1.it (D.P.)
3. Department of Industrial Chemistry, University of Bologna, 4 Viale del Risorgimento, 40136 Bologna, Italy; aisha.matayeva2@unibo.it (A.M.); andrea.fasolini2@unibo.it (A.F.); angelo.vaccari@unibo.it (A.V.); f.basile@unibo.it (F.B.); giuseppe.fornasari@unibo.it (G.F.)
* Correspondence: miky@unive.it

Received: 22 June 2020; Accepted: 10 July 2020; Published: 15 July 2020

Abstract: The transformation of glucose into valuable biochemicals was carried out on different MCM-41-supported metallic and bimetallic (Co, Co-Fe, Co-Mn, Co-Mo) catalysts and under different reaction conditions (150 °C, 3 h; 200 °C, 0.5 h; 250 °C, 0.5 h). All catalysts were characterized using N_2 physisorption, Temperature Programmed Reduction (TPR), Raman, X-ray Diffraction (XRD) and Temperature Programmed Desorption (TPD) techniques. According to the N_2-physisorption results, a high surface area and mesoporous structure of the support were appropriate for metal dispersion, reactant diffusion and the formation of bioproducts. Reaction conditions, bimetals synergetic effects and the amount and strength of catalyst acid sites were the key factors affecting the catalytic activity and biochemical selectivity. Sever reaction conditions including high temperature and high catalyst acidity led to the formation mainly of solid humins. The NH_3-TPD results demonstrated the alteration of acidity in different bimetallic catalysts. The $10Fe10CoSiO_2$ catalyst (MCM-41 supported 10 wt.%Fe, 10 wt.%Co) possessing weak acid sites displayed the best catalytic activity with the highest carbon balance and desired product selectivity in mild reaction condition. Valuable biochemicals such as fructose, levulinic acid, ethanol and hydroxyacetone were formed over this catalyst.

Keywords: glucose; biochemicals; MCM-41; bimetallic; reactivity; product selectivity

1. Introduction

The depletion of fossil fuels along with the environmental problems associated with their utilization promote new processes for the generation of fuels based on renewable sources [1]. The use of biomass, typically lignocellulosic biomass, in the production of fuels, fuel additives or added-value chemicals has attracted considerable interest, becoming a potential research area [2]. Biomass can be converted into biofuels and valuable chemicals via chemical or thermochemical processes; among them, aqueous phase reaction is an effective method to convert lignocellulose into biochemicals [3]. Lignocellulosic biomass is principally constituted of cellulose, hemicellulose and lignin [4,5]. Due to the complex nature of biomasses and their chemical compositions, they have different reactivities, and conversion processes can occur via various reaction pathways. Thus, researchers often prefer to use typical model components focusing on special reaction stages in order to study the conversion process [6,7]. Glucose, the most plentiful and approachable monosaccharide unit in the lignocellulosic biomass, is the most desirable feedstock for the production of valuable biochemicals [8]. One of the most significant reaction pathways is the isomerization of glucose to fructose, followed by dehydration of fructose to 5-hydroxymethylfurfural (5-HMF) and rehydration of 5-HMF to levulinic acid and formic acid [9]. Moreover, retro aldol condensation may occur with glucose, yielding glycolaldehyde and erythrose, or with fructose to give dihydroxyacetone and glyceraldehyde. Glyceraldehyde can be hydrogenated to glycerol or rearranged to lactic acid, which can be further hydrogenated with propionic acid or decarboxylated to yield ethanol. Another pathway involving C-O and C-C cleavage leads to acetic acid, which can be obtained from glycerol [10–14].

The competing reaction pathways happening through biomass conversion lead to different bioproducts with relatively low yields and difficulties in separation. It is important to underline one of the major drawbacks of working with sugars: the formation of humins, i.e., insoluble, heavy compounds that form under glucose transformation conditions. Therefore, catalysts and the reaction conditions play a crucial role in the control of reactions or the production of the desirable bioproduct, higher feedstock conversion and also in avoiding byproducts [15]. Compared to homogeneous catalysts, heterogeneous catalysts can be used in viable greener methods and approaches for efficient biomass transformation. High thermal and mechanical stability, recovery and recyclability are the main advantages of heterogenous catalysts [16]. In recent years, a wide variety of heterogenous catalysts has been utilized for biomass transformation, such as carbon-based materials, mesoporous silica, zeolites, metal oxide supported metals, organic polymers and ionic liquids [17,18].

In particular, mesoporous silica materials such as SBA-15, KIT-6, and MCM-41 have been investigated as supports owing to their high surface area and large pore volume, flexible and tunable properties, and easy diffusion of large molecules, enabling efficient transformation [2,7,19]. For instance, Qing Xu et al. reported the effect of a Sn-containing silica mesoporous framework (Sn-MCM-41) on the conversion of glucose to 5-HMF in ionic liquid. The results showed that Sn can act as a highly active Lewis acid center in conjunction with the silanol group of MCM-41, which catalyzes both the isomerization of glucose into fructose and the dehydration of fructose to HMF without the addition of a mineral Brønsted acid catalyst [20]. Other authors reported the use of hybrid catalysts such as $CrCl_3$ and HY zeolite for the production of lactic acid (LA) acid from glucose, with better performance compared to parent HY catalyst because of their higher acidity [21,22]. Cao Xuefei et al. used various transition metal sulfates (Mn^{2+}, Fe^{2+}, Fe^{3+}, Co^{2+}, Ni^{2+}, Cu^{2+}, and Zn^{2+}) to achieve glucose, fructose and cellulose conversion into several chemicals; the authors pointed out that different metal salts showed different reactivities regarding the conversion of sugars. Among these metal ions, Zn^{2+} and Ni^{2+} were more selective towards LA, whereas Cu^{2+} and Fe^{3+} showed high levels of efficiency for the conversion of glucose and cellulose into LA and formic acid at high temperature [23].

Materials possessing high surface areas and large pore sizes with more accessible acidic moieties are crucial for catalyst preparation. To the best of our knowledge, no studies have reported the use of mesoporous-supported, bimetallic catalysts for the aqueous phase transformation of glucose into biochemicals. Furthermore, using metals such as Fe, Co, Mo and Mn, which are cheaper and more widely available, as the active phase may be interesting and of crucial importance.

Here, we report on four catalysts including, i.e., Co, Co-Mn, Co-Mo, and Co-Fe supported MCM-41 (hexagonal mesoporous silica) for the aqueous phase formation of valuable chemicals from glucose under different reaction conditions. The aim of this study is to benefit from the high surface area, large pore size and acidity of the silanol group in MCM-41 and the metal active phase in order to facilitate the penetration of substrates into the catalyst pores.

2. Experimental Part

2.1. Catalysts Synthesis

A silica mesoporous support (MCM-41) was synthesized following the procedure described by Ghedini et al. [24]. Hexadecyltrimethylammonium bromide (CTAB, Aldrich) was first dissolved in a NaOH aqueous solution at room temperature (r.t) under stirring; then, the required amount of tetraethyl orthosilicate (TEOS) was added. The resulting mixture was aged in an autoclave at 150 °C for 22 h, and thereafter filtrated, thoroughly washed and dried at room temperature. The surfactant was removed by calcination at 500 °C for 6 h in air flow (50 mL/min).

For the monometallic catalyst, the active phase was introduced on a silica support by incipient wetness impregnation using an aqueous solution of $Co(NO_3)_2 \cdot 6H_2O$ in order to obtain 20 wt.% of metal loading.

For bimetallic samples, the precursors were introduced by co-impregnation of the previous Co solution and the corresponding precursors including $(NH_4)_6Mo_7O_{24} \cdot 4H_2O$, $MnSO_4 \cdot H_2O$ and $Fe(NO_3)_3 \cdot 9H_2O$) in order to obtain a nominal value of 10 wt% for each metal. Finally, the samples were dried and calcined at 500 °C in air flow (50 mL/min) for 6 h.

The catalysts were labelled: $20CoSiO_2$, $10Mo10CoSiO_2$, $10Mn10CoSiO_2$ and $10Fe10CoSiO_2$.

2.2. Catalysts Characterisation

2.2.1. Nitrogen Physisorption

Nitrogen physisorption measurements were performed at −196 °C using a Micromeritics Tristar II Plus sorptometer (MICROMERITICS, Norcross, GA, USA). The sample (~400 mg) was outgassed at 200 °C for 2 h in vacuum prior to the sorption experiment. The surface area was calculated using the BET equation [25], and the total pore volume, V_{tot}, was measured as the adsorbed amount of N_2 at P/P_0 values near 0.98. Pore size distribution was determined by the BJH method [26] applied to the N_2 adsorption isotherm branch [27,28].

2.2.2. Temperature Programmed Reduction (TPR)

TPR measurements were carried out with a lab-made instrument at CATMAT laboratory, Ca' Foscari University of Venice. The analysis was performed under 5% H_2/Ar (40 mL/min) from 25 °C to 800 °C with a heating rate of 10 °C/min. The H_2 consumption was analyzed by a Micrometrics TPD-TPR 2900 analyzer equipped with a TCD detector (Gow-Mac 24-550 TCD instrument CO, Bethlehem, PA, USA).

2.2.3. X-ray Powder Diffraction (XRD)

A XRD (PW1769, Philips Analytical, Eindhoven, The Netherlands) using Cu-Kα (Ni-filtered) radiation was used for crystalline phase determination. The measured 2θ angle range was 10.0°–70.0° with a step size of 0.02° and a counting time of 1.25 s per step. The size of the metal particle phase was obtained using the Scherrer equation [26]. The correction for instrument broadening was applied after background subtraction and curve-fitting procedures on the assumption of Lorentzian peak profiles.

2.2.4. Raman Spectroscopy

Raman spectra were collected on powder samples at room temperature in back-scattering geometry using an inVia Renishaw 1000 spectrometer equipped with an air-cooled, charge-coupled device (CCD) detector and edge filters. A 488.0 nm emission line from an Ar^+ laser was focused on the sample using a Leica DLML microscope with 5 × or 20 × objectives and an incident beam power of about 5 mW. A solid-state laser emitting at 785 nm with low power to avoid sample damage (about 2 mW) was used to analyze the $10Fe10CoSiO_2$ catalyst. The spectra were calibrated using the 520.5 cm^{-1} line of a silicon wafer. The spectral resolution was 3 cm^{-1}. Data analyses included baseline removal and curve fitting using the Gauss Lorentzian cross-product function in the Peakfit 4.12 software (Systat Software Inc., San Jose, CA, USA, 2007).

2.2.5. Temperature-programmed Desorption (NH_3-TPD)

NH_3-TPD analyses of samples were carried out using lab-made equipment at CATMAT laboratory, Ca' Foscari University of Venice in order to study the acidity of the catalysts. First, 100 mg of the catalyst was charged in a quartz reactor and degassed in He with a flow rate of 40 mL/min at 500 °C for 90 min. The catalyst was then cooled to room temperature (25 °C) prior to adsorption of ammonia. Then, the adsorption of 5% NH_3/He with a flow rate of 40 mL/min at 25 °C for 30 min was performed. The physisorbed ammonia was removed from the catalyst surface by passing He (40 mL/min) at room temperature for 10 min. The desorption profile of NH_3-TPD was recorded using a Micrometrics TPD-TPR 2900 analyzer equipped with a thermal conductivity detector TCD (Gow-Mac 24-550 TCD instrument CO, Bethlehem, PA, USA) from 25 to 1000 °C at a heating rate of 10 °C/min under the flow of He (40 mL/min).

2.3. Aqueous Phase Transformation Catalytic Tests

Aqueous phase reforming (APR) tests were carried out in a 300 mL stainless-steel Parr autoclave loaded with a 0.3–3.0 wt.% solution of glucose in water and 0.45 g of catalyst. All the catalysts were pelletized and reduced at 500 °C for 3 h under a 10 % (v/v) H_2/N_2 flow before each test.

The experiments were performed by placing the catalyst and glucose water solution (50 mL) into the autoclave. Thereafter, the sealed autoclave was first purged under N_2 flow for a few minutes to remove oxygen in the gas phase, and then heated to the desired temperature at 4.2 °C/min. When the desired temperature was reached, the reaction was started. The heating period was not considered in the reaction time. All the reactions were performed in a temperature range of 150–250 °C at autogenous pressure for different durations. At 150 °C, the tests were conducted for 3 h to allow glucose conversion and product formation, while at higher temperatures, the reaction was carried out for 0.5 h to avoid complete glucose transformation into humins. Considering the heating step, the following reaction times were employed: 3.5 h at 150 °C (0.5 h heating); 1.25 h at 200 °C (0.75 h heating); and 1.5 h at 250 °C (1,0 h heating). At the end of the reaction time, the autoclave was quenched in ice and allowed to cool to room temperature over 30–40 min. The reaction mixture was analyzed using Agilent HPLC over a Rezex ROA Organic Acid column (0.0025 M H_2SO_4 eluent, oven temperature 60 °C and 0.6 mL/min flux) with a RID detector. Gas analyses were performed in an off-line Thermo Focus GC with a carbon molecular sieve column (Carbosphere 80/100 6 * 1/8) and TCD detector.

The glucose conversion, carbon balance and product yields were calculated using the following equations:

$$\text{Conversion (\%)} = \frac{(\text{mmol sub in}) - (\text{mmol sub out})}{\text{mmol sub in}} \cdot 100 \qquad (1)$$

$$\text{Carbon balance (\%)} = \frac{\sum_i (\text{mmol out}) \cdot (\text{C atoms})}{(\text{mmol subin} \cdot \text{C atoms glc})} \cdot 100 \qquad (2)$$

$$\text{Yield (\%)} = \frac{\text{mmol i out}}{\text{mmol sub in}} \cdot 100 \qquad (3)$$

where i represents the general product of the reaction.

A couple of randomly chosen center-points were duplicated according to a statistical approach to estimate the variability of the results. The maximum standard deviation exceeded 0.03 and 3 for the conversion and the carbon balance, respectively.

3. Result and Discussion

3.1. Catalysts Characterization

A TEM image of MCM-41 as a catalyst support is shown in Figure 1. It demonstrates the presence of a highly ordered array and layered structure.

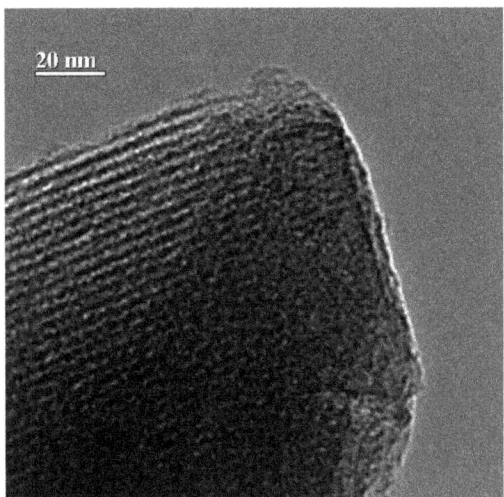

Figure 1. TEM image of MCM-41.

The X-ray powder diffraction pattern of the 20CoSiO$_2$ sample (Figure S1) shows peaks at 2θ values of 31.5, 36.9, 38.6, 44.8, 59.4, 65.2° corresponding to the Co$_3$O$_4$ (220), (311), (222), (400), (511), (440) planes, respectively [JCPDS card 9-418]. For the 10Fe10Co SiO$_2$ sample, the peaks at 2θ values of 35.4 and 36.7 denote the most intense lines of the Fe$_3$O$_4$ (311) and of Co$_3$O$_4$(311) phases, respectively [JCPDS card 19-629; JCPDS card 9-418], whereas features at 2θ values of about 31.0, 58.0 and 64 are hardly distinguishable.

The 10Mn10Co SiO$_2$ sample exhibits a broad peak at 2θ of 36.6° and barely detectable features at 2θ of 59° and 65° which were assigned to the most intense Co$_3$O$_4$ (311), (511), (440) planes; no peaks of manganese-containing phases were detected. The average crystallite size of Co$_3$O$_4$, determined by Scherrer equation (Lorentzian peak profile; 2θ, 36.6°), was ~30 nm for the 20CoSiO$_2$ sample, ~ 10 nm for the 10Fe10CoSiO$_2$ sample and ~ 6 nm for the 10Mn10CoSiO$_2$ sample. An average crystallite size of about 10 nm was determined for Fe$_3$O$_4$ (2θ, 35.4°) in the 10Fe10CoSiO$_2$ sample. These results indicate that in the bimetallic catalysts, the spreading of surface species was favored, with an ensuing reduction in material crystallinity [29].

As for the 10Mo10CoSiO$_2$ system, the XRD pattern shows reflections due to MoO$_3$ [JCPDS card 5-0508] and to CoMoO$_4$, identified by the peaks at 2θ values of 26.4 and 31.9° and 35.5° [JCPDS card 21-868]. The average crystallite size results of ~100 nm for MoO$_3$ and ~50 nm for CoMoO$_4$ are indicative of highly crystalline materials [29]

Raman Spectroscopy was applied to obtain information concerning the chemical structure and molecular interactions among the various components in both monometallic and bimetallic samples (Figure 2A).

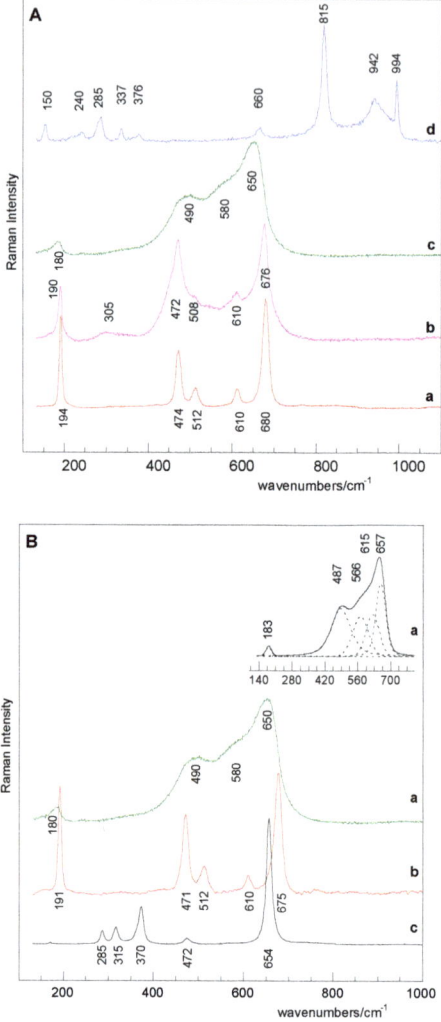

Figure 2. (**A**) Raman spectra of 20CoSiO$_2$ (a), 10Fe10CoSiO$_2$ (b) 10Mn10CoSiO$_2$ (c) and 10Mo10CoSiO$_2$ (d); (**B**) Raman spectra of 10Mn10CoSiO$_2$ (a), Co$_3$O$_4$ (b) and Mn$_3$O$_4$ (c) oxides. inset: curve fitting results obtained for the 10Mn10CoSiO$_2$ sample.

The 20CoSiO$_2$ sample (Figure 2A, curve a) exhibits the sharp Raman-active modes (F_{2g}, E_g, F_{2g}, F_{2g} and A_{1g}, respectively) predicted for the Co$_3$O$_4$ spinel structure [30], confirming the XRD results. Co$_3$O$_4$ has a normal spinel structure with Co^{2+} positioned at the tetrahedral site and Co^{3+} at octahedral site. The most intense mode (A_{1g}) is attributed to the octahedral site symmetry, whereas the weakest modes (F_{2g} and E_g) are related to the combined vibrations of tetrahedral sites and octahedral oxygen motions [31]. In the 10Fe10CoSiO$_2$ spectrum (Figure 2A, curve b), the most intense bands (190, 472 and 676 cm^{-1}) are assigned to Co$_3$O$_4$ modes [24], while the broad and low intensity one at about 305 cm^{-1} clearly identifies Fe$_3$O$_4$ nanoparticles [25], in line with XRD analysis. The presence of both Fe$_3$O$_4$ and Co$_3$O$_4$ is further confirmed by the broad features at 508 and 610 cm^{-1} and the asymmetric shape of the main band at 676 cm^{-1}, resulting from the superimposition of some of their bands.

The bimetallic 10Mn10CoSiO$_2$ sample (Figure 2A), curve c) showed a broad unresolved feature with some prominent components (at about 180, 490, 580 and 650 cm^{-1}), indicative of nanostructured surface species [32].

For the 10Mo10CoSiO$_2$ sample (Figure 2A), curve d), bands characteristic of the α-MoO$_3$ crystal phase [33] (sharp peaks at about 240, 285, 337, 376, 660, 815 and 994 cm^{-1}) and of the CoMoO$_4$ structure [34] (broad feature at about 942 cm^{-1}) were identified. As for the α-MoO$_3$ phase, the narrow peak at 994 cm^{-1} could be assigned to the terminal oxygen (Mo = O) stretching mode, and the peaks at 811 cm^{-1} and at 660 cm^{-1} were attributed to the doubly (Mo$_2$-O) and triply coordinated (Mo$_3$-O) oxygen stretching mode, respectively, whereas the low intensity of peaks in the 200–400 cm^{-1} region were due to the Mo–O bending modes.

In Figure 2B, the broad feature representing the 10Mn10CoSiO$_2$ sample (curve a) is inspected with reference to the spectral features of Co$_3$O$_4$ (curve b) and Mn$_3$O$_4$ (curve c) oxides. Both the oxides showed sharp peaks indicative of a crystalline structure, with the Mn$_3$O$_4$ spectrum being characterized by an intense peak at 654 cm^{-1} assigned to Mn-O vibrations of manganese (II) ions in tetrahedral coordination [35]. With reference to the Co$_3$O$_4$ spectrum, the addition of manganese species caused bands shift to a lower frequency, broadening and coalescence of some of the vibration modes in the 500–700 cm^{-1} region. These changes could have arisen from the formation of nanostructured species, which caused changes in the coordination and local symmetry of the pure oxide components [32,36]. Bands at 566 and 615 cm^{-1} identified by Curve fitting results (Figure 2B), curve a inset) suggested that surface Mn$_3$O$_4$ and Co-Mn mixed oxide species had formed [37,38], as also supported by bands at 487 cm^{-1} and at 657 cm^{-1}, arising from coalescence (bands at 471 and 512 cm^{-1}) and down shift of Co$_3$O$_4$ modes.

In order to investigate the specific surface areas and pore size distribution of the catalysts, nitrogen physisorption was performed. The adsorption–desorption profile of pristine silica support exhibited a type IV isotherm, which was typical of a high surface area mesoporous material (Figure S2 black line), in accordance with IUPAC classification [28]. The BET surface area of the obtained material was 1000 m^2/g and the pore volume of 0.4 cm^3/g. A high surface area and the mesoporous structure of a catalyst have direct and indirect effects on the reaction results, i.e., increasing the active metal dispersion on the catalyst surface, which could make the active phase more accessible for the reactant and improve the activity of the catalyst. Moreover, a mesoporous catalyst structure made the reactant diffusion and product formation more efficient.

The adsorption–desorption profiles of Co-supported catalysts (Figure S2) were similar to the isotherm profile of SiO$_2$, proving that the pristine support kept its structure after the deposition of the metal phase. However, the isotherm and surface area values of the 10Mo10CoSiO$_2$ sample substantially changed compared to the support ones. The textural properties, including surface areas and pore volumes, decreased somewhat (Table 1) with the deposition of metal active phase. As shown in Table 1, the 10Mo10CoSiO$_2$ catalyst presented the lowest surface area; this was ascribed to the deposition of a large number of cobalt species inside the pores of the mesoporous silica structure. Moreover, according to the XRD analysis, the sample consisted of large crystallites of MoO$_3$ and CoMoO$_4$, leading to a decrease in specific surface area due to pore blocking.

Table 1. Surface area and pore size of catalysts.

Catalyst.	Specific Surface Area (m^2/g)	Average Pore Diameter (nm)	Pore Volume (cm^3/g)
SiO$_2$	1000	-	0.40
20CoSiO$_2$	460	1.6	0.32
10Fe10CoSiO$_2$	510	1.5	0.33
10Mn10CoSiO$_2$	403	1.6	0.30
10Mo10CoSiO$_2$	185	2.9	0.23

A H_2-TPR analysis was carried out to determine the reducibility of metal species and the possible interaction between the metals present on each catalyst. Figure 3 shows the profiles related to hydrogen consumption as a function of temperature for mono and bimetallic Co-based samples.

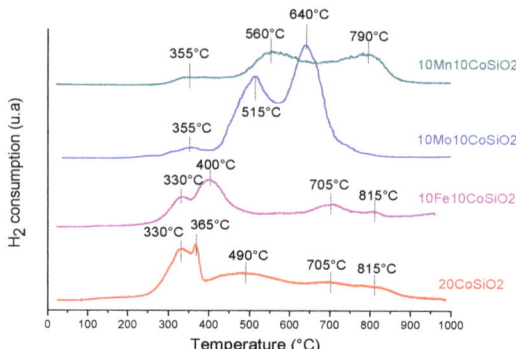

Figure 3. TPR profiles of the mono and bimetallic Co-based samples.

The TPR profiles of the catalysts showed several peaks indicating the formation of different Cobalt/Promoters species and their various interactions with the support. The pattern of the monometallic $20CoSiO_2$ sample presented two major reduction peaks at 330 °C and 365 °C. The peak at a lower temperature (330 °C) was assigned to the reduction of Co_3O_4 species to CoO, while the second one (365 °C) was ascribed to the reduction of CoO to Co^0 [39–41]. The presence of a broad feature situated between 400 °C and 850 °C could instead be related to the reduction of cobalt species strongly interacting with the support [40].

The TPR profile of the $10Fe10CoSiO_2$ catalyst was similar to that of the $20CoSiO_2$ sample, even though the position of the second peak was slightly shifted toward a higher temperature (from 365 °C to 400 °C) and the broad feature centered at 490 °C disappeared, indicating stronger interaction with the support. In addition, the second peak at 400 °C could be ascribed to the reduction of both CoO to Co^0 and Fe_3O_4 to Fe^0 [42].

The TPR profile of the $10Mo10CoSiO_2$ catalyst showed two well-resolved reduction peaks centered at 515° and 640 °C, and a low intensity broad peak at 355 °C. With regards to the peak at 515 °C, it was not possible to assign it to specific metal species, since it may have been due to the reduction step of CoO to Co^0 shifting toward higher temperatures, or to the reduction of MoO_3 to MoO_2 which occurs over a temperature range of 450–650 °C [43]. On the other hand, the peak at 650°, which was not present in TPR profile of the monometallic catalysts ($20CoSiO_2$), could be ascribed to the reduction of MoO_2 to Mo^0 [44], and to the reduction of $CoMnO_4$ [45].

Finally, compared to the $10CoSiO_2$ catalyst, the presence of Mn in the $10Mn10CoSiO_2$ catalyst led to a shift to higher temperature of the first reduction peak, which was related to the harder reduction of $Co_3O_4 \rightarrow Co^0$; this was proven by using less hydrogen. However, the reduction profile of the $10Mn10CoSiO_2$ sample showed broad peaks at 560° and 790 °C, that could be related to the reduction of Mn^{4+} and Mn^{3+} to Mn^{2+} [37], and to the reduction of Co species, which have strong metal–support interactions, and Co–Mn mixed oxide species, as also identified by Raman analysis [46,47].

NH_3-TPD was performed in order to study the acidic features of the catalysts. The NH_3-TPD spectra of all the catalysts are presented in Figure 4. Two peaks in two temperature ranges were found in the spectra of the catalysts, indicating the presence of two types of acidity. The peak in the temperature range of around 100–150 °C was associated with the week acidic sites, while that in the temperature range of approximately 700–850 °C was ascribed to the strong acidic sites [48]. $20CoSiO_2$ showed a small peak in a higher temperature range (750–850 °C), indicating the presence of slight amounts of strong acidic sites, probably due to the acidity of SiO_2 and the formation of CoO_x over the support. Compared to the Co monometallic catalyst, mixed Fe-Co and Mo-Co phases exhibited weak acid sites, while the mixed Mn-Co phase increased the intensity of the strong acidity peak. The various bimetallic phases showed different acidic properties [48–50].

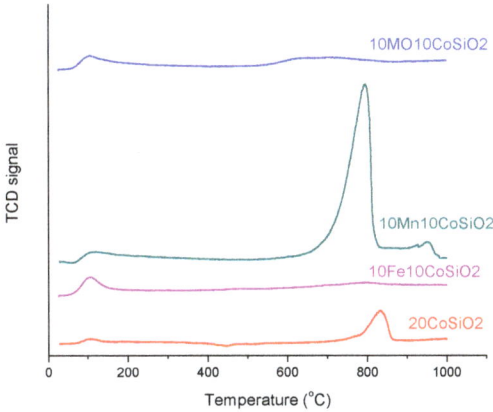

Figure 4. NH3-TPD profiles for synthesized catalysts.

3.2. Catalytic Test

In order to understand the reactivity of glucose under different reaction conditions, and to determine the effect of the catalysts, blank test were performed. The results obtained at different reaction conditions (150 °C for 3.0 h, 200 °C and 250 °C for 0.5 h) are shown in Figures 5 and 6 and Table 2.

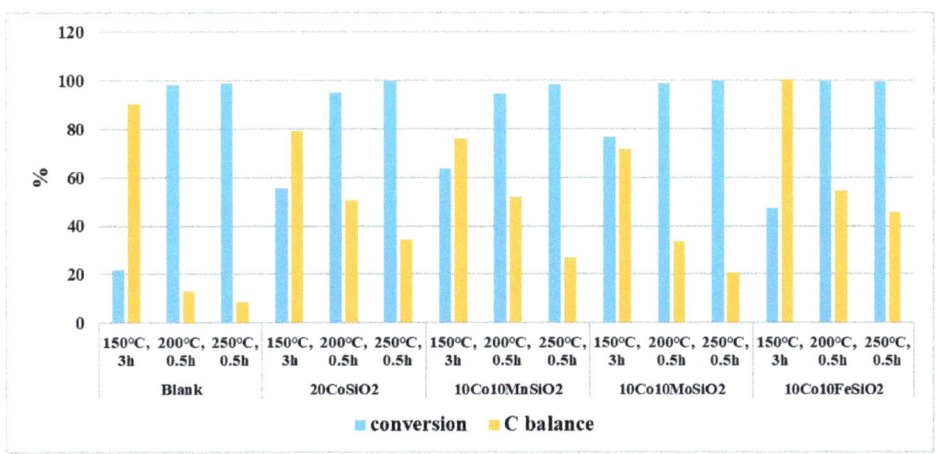

Figure 5. Glucose conversion and carbon balance in the presence of catalysts.

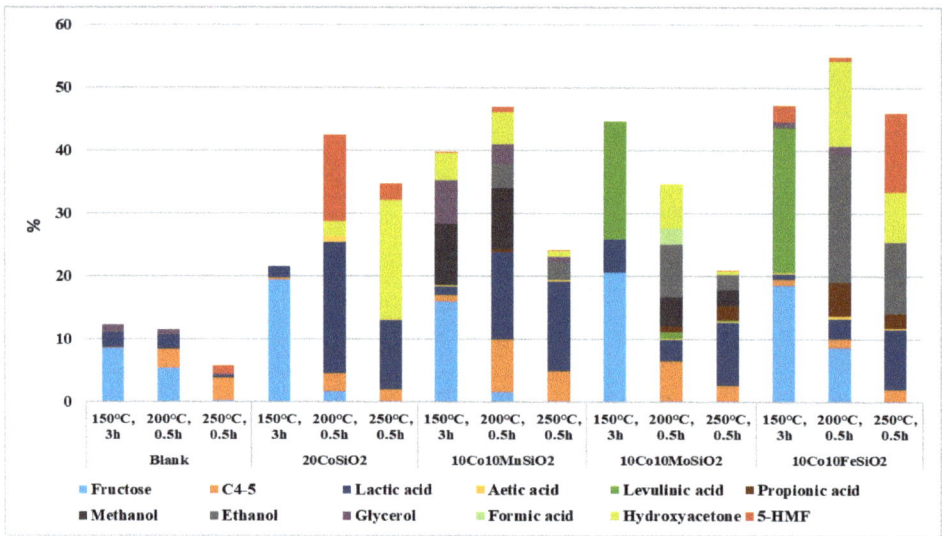

Figure 6. Product distribution for the transformation of glucose in the presence of catalysts.

The conversion of glucose without any catalyst was only 22% at 150 °C (Figure 5). Fructose, C4-C5 sugars, glycerol and lactic acid were also detected. This suggests that the isomerization, retro-aldol condensation and dehydration reactions occurred under reaction conditions without any catalyst. Further increases in temperature up to 200 and 250 °C resulted in the almost complete conversion of glucose (98 and 99%, respectively). However, the yields of the identified products did not increase appreciably. The obtained data evidenced high conversion and carbon loss at temperatures above 200 °C due to changes in the reaction pathways; at high temperatures, glucose was converted into both insoluble humins and soluble polymeric byproducts [51]. This could be attributed to the oligomerization reaction of both glucoses, HMF and other reaction products (fructose, other C4-C5 sugars). This underlines one of the main drawbacks of working with sugars [12]. For this reason, different catalysts under the same reaction conditions were screened to identify a reliable method to reduce these byproducts.

In the presence of 20Co/SiO$_2$, even at a low reaction temperature (150 °C), the conversion of glucose was 58%, and fructose was identified as the main product according to NH$_3$-TPD tests that showed the presence of slight number of strong acidic sites active in the isomerization of glucose to fructose. However, the temperature was too low to allow further conversion to occur of fructose to other products. Hence, at 200 °C, the presence of Co nearly completely converted the glucose, and the yield of lactic acid increased to 19%, along with larger amounts of HMF (14%) from the dehydration of fructose. As shown in the experimental data, it seems that the Co-based catalyst with few strong acid sites according to NH$_3$-TPD analyses favored dehydration and the breakage of C–C bonds, in comparison to the blank experiment [52,53].

However, by further increasing the temperature to 250 °C, the yield sum was lower, with hydroxyacetone being the main product, followed by lactic acid. This caused a decrease in the carbon balance values, together with an increase in conversion due to the formation of humins at higher temperature.

Table 2. Glucose conversion, product selectivity and carbon balance obtained under different conditions.

Catalyst	Reaction Conditions	Conversion, %	Fructose	C4-5	Lactic Acid	Acetic Acid	Levulinic Acid	Propionic Acid	Methanol	Etanol	Glycerol	Formic Acid	Hydroxyacetone	5-HMF	Carbon Balance*, %
No cat	150° C, 3h	22	40	<1	11	-	-	-	-	-	5	-	-	-	91
	200° C, 0.5h	98	5	3	2	-	-	-	-	-	<1	-	-	-	13
	250 °C, 0.5h	99	<1	4	<1	0.0	-	-	-	-	<1	-	-	1	9
Co	150 °C, 3 h	56	35	<1	3	-	-	-	-	-	-	-	-	-	79
	200 °C, 0.5h	95	2	3	22	1	-	-	-	-	-	-	3	14	51
	250 °C, 0.5 h	100	-	2	11	<1	-	-	-	-	-	-	19	3	35
Co-Mn	150 °C, 3 h	64	25	2	2	<1	-	-	15	-	11	-	7	<1	76
	200 °C, 0.5 h	95	2	9	15	-	<1	10	4	3	-	-	5	<1	52
	250 °C, 0.5h	99	-	5	14	<1	-	-	3	<1	-	-	<1	<1	27
Co-Mo	150 °C, 3 h	77	27	-	7	-	25	-	-	-	-	-	-	-	-
	200 °C, 0.5h	99	-	6	3	<1	1	-	5	9	-	3	7	-	-
	250 °C, 0.5h	100	-	2	10	<1	<1	2	3	2	-	-	<1	-	-
Co-Fe	150 °C, 3 h	48	39	2	2	<1	48	-	-	1	<1	-	-	5	100
	200 °C, 0.5h	100	9	1	3	<1	-	5	-	20	1	-	14	<1	55
	250 °C, 0.5 h	100	-	20	9	<1	-	2	-	11	-	-	8	12	46

Among the tested bimetallic catalysts, the best conversion was achieved over 10Mo10CoSiO$_2$ (77%), followed by 10Mn10CoSiO$_2$ (64%) at 150 °C. Compared to the Co monometallic catalyst, 10Mn10CoSiO$_2$ with strong acidity enhanced the C–C bond cleavage, facilitating the generation of C3 products (lactic acid, hydroxyacetone, glycerol) [54]. In contrast, the 10Fe10CoSiO$_2$ catalyst with weak acidity exhibited the lowest activity for glucose conversion (48%). Thus, it was confirmed that the presence, amount and strength of acid sites are important variables in determining the extent of glucose conversion into biochemicals. This notwithstanding, the use of 10Fe10CoSiO$_2$ resulted in a complete carbon balance, thereby inhibiting the pathway to the formation of humins due to the slight acidity of the catalyst. It was demonstrated that strong acidity can increase product polymerization and the production of humins, and therefore, cause a decrease in product yields and carbon balance. This effect was also demonstrated by the formation of levulinic acid for both 10Fe10CoSiO$_2$ and 10Mo10CoSiO$_2$ catalysts with weak acid sites [54,55].

At higher temperatures (200 and 250 °C), 10Mo10CoSiO$_2$ showed lower performance in terms of glucose conversion and sum yields of products compared to 20CoSiO$_2$ only.

Even though the substitution by Mn or Fe did not significantly affect the conversion of glucose and the carbon balance at 200 °C compared to those obtained using only Co, different reaction product distributions occurred. For instance, the presence of Mn promoted the formation of a wide range C3 products, C4-C5 sugars and methanol, thus enhancing C-C cleavage and dehydration. In comparison, 10Fe10CoSiO$_2$ promoted the production of ethanol and hydroxyacetone. This might stem from the synergistic effect between the two types of metal species and the difference of acidity in the two catalysts [54,56].

Completely different behavior was observed with the bimetallic catalysts at 250 °C. The data obtained from the experiment using 10Mn10CoSiO$_2$ show the lowest carbon balance among the tested catalysts. Considering the reaction product distribution, it seems that the formation of humins could be due to the condensation reaction of C4-C5 sugars, one of the main products at 200 °C, under harsh reaction conditions, as demonstrated in NH$_3$-TPD results.

The best balance (46%) was obtained at 250 °C, with complete conversion using the 10Fe10CoSiO$_2$ catalyst, owing to its lower acidity, as shown by NH$_3$-TPD tests. The main product was 5-HMF, due to the dehydration of fructose. Ethanol was produced with a yield of 11%, while the yield of lactic acid reached 9%.

4. Conclusions

MCM-41-supported Co and bimetallic Co-Mn, Co-Mo and Co-Fe catalysts were investigated regarding the aqueous phase transformation of glucose into biochemicals, operating under different reaction conditions. With increases in temperature and the number and strength of catalyst acid sites, the conversion of glucose increased, but resulted in a low carbon balance due to the formation of humins. Isomerization was the predominant reaction at lower temperatures, while humins prevailed at higher temperatures, longer reaction times and higher catalyst acidity. The synergy of Fe, Mn and Mo with the Co increased the activity at 150 and 200 °C, while 250 °C hydrothermal conditions favored the retro aldol condensation reaction of glucose and its intermediates (fructose, C4-C5 sugars and 5-HMF) to form humins. The best catalytic reactivity was obtained under mild reaction conditions with the weak acidic sites of the 10Fe10CoSiO$_2$ catalyst, yielding valuable biochemicals such as fructose, levulinic acid, ethanol and hydroxyacetone.

Supplementary Materials: The following are available online at http://www.mdpi.com/2227-9717/8/7/843/s1, Figure S1: XRD patterns of the 20CoSiO$_2$ (a), 10Fe$_{10}$CoSiO$_2$ (b) 10Mn$_{10}$CoSiO$_2$ (c) and 10Mo$_{10}$CoSiO$_2$ (d) samples*Co$_3$O$_4$; ^Fe$_3$O$_4$; MoO$_3$; #CoMoO$_4$, Table S1: N$_2$ adsorption/desorption isotherms of catalysts and pristine support.

Author Contributions: Conceptualization, S.T.; methodology, D.P. and E.G.; validation, F.M., F.B. and D.G.; formal analysis and investigation, S.T., A.M. and A.F.; writing—original draft preparation, S.T.; supervision, M.S.; project administration, G.F.; funding acquisition, A.V. All authors were contributed in the manuscript writing. All authors have read and agreed to the published version of the manuscript.

Funding: This research was funded by Italian National Research Program PRIN 2015 "Heterogeneous Robust catalysts to upgrade low value biomass stream (HERCULES)" project.

Conflicts of Interest: The authors declare no conflicts of interest.

References

1. Wang, J.; Xi, J.; Wang, Y. Recent advances in the catalytic production of glucose from lignocellulosic biomass. *Green Chem.* **2015**, *17*, 737–751. [CrossRef]
2. Chen, S.; Maneerung, T.; Tsang, D.C.; Ok, Y.S.; Wang, C.-H. Valorization of biomass to hydroxymethylfurfural, levulinic acid, and fatty acid methyl ester by heterogeneous catalysts. *Chem. Eng. J.* **2017**, *328*, 246–273. [CrossRef]
3. Kang, S.; Fu, J.; Zhang, G. From lignocellulosic biomass to levulinic acid: A review on acid-catalyzed hydrolysis. *Renew. Sustain. Energy Rev.* **2018**, *94*, 340–362. [CrossRef]
4. Somerville, C.; Youngs, H.; Taylor, C.; Davis, S.C.; Long, S. Feedstocks for Lignocellulosic Biofuels. *Science* **2010**, *329*, 790–792. [CrossRef] [PubMed]
5. Sun, Y.; Lu, X.; Zhang, S.; Zhang, R.; Wang, X. Kinetic study for Fe (NO3)3 catalyzed hemicellulose hydrolysis of different corn stover silages. *Bioresour. Technol.* **2011**, *102*, 2936–2942. [CrossRef]
6. Chen, H. *Biotechnology of Lignocellulose*; Springer: Berlin/Heidelberg, Germany, 2014.
7. Karnjanakom, S.; Guan, G.; Asep, B.; Hao, X.; Kongparakul, S.; Samart, C.; Abudula, A. Catalytic Upgrading of Bio-Oil over Cu/MCM-41 and Cu/KIT-6 Prepared by β-Cyclodextrin-Assisted Coimpregnation Method. *J. Phys. Chem. C* **2016**, *120*, 3396–3407. [CrossRef]
8. Qib, X.; Watanabe, M.; Aida, T.M.; Smith, R.L. Fast Transformation of Glucose and Di-/Polysaccharides into 5-Hydroxymethylfurfural by Microwave Heating in an Ionic Liquid/Catalyst System. *ChemSusChem* **2010**, *3*, 1071–1077. [CrossRef]
9. Choudhary, V.; Pinar, A.B.; Sandler, S.I.; Vlachos, D.G.; Lobo, R.F. Xylose Isomerization to Xylulose and its Dehydration to Furfural in Aqueous Media. *ACS Catal.* **2011**, *1*, 1724–1728. [CrossRef]
10. Tanksale, A.; Beltramini, J.N.; Lu, G.Q. Reaction Mechanisms for Renewable Hydrogen from Liquid Phase Reforming of Sugar Compounds. *Dev. Chem. Eng. Miner. Process.* **2008**, *14*, 9–18. [CrossRef]
11. Fasolini, A.; Cespi, D.; Tabanelli, T.; Cucciniello, R.; Cavani, F. Hydrogen from Renewables: A Case Study of Glycerol Reforming. *Catalysts* **2019**, *9*, 722. [CrossRef]
12. Sasaki, M.; Goto, K.; Tajima, K.; Adschiri, T.; Arai, K. Rapid and selective retro-aldol condensation of glucose to glycolaldehyde in supercritical water. *Green Chem.* **2002**, *4*, 285–287. [CrossRef]
13. Fasolini, A.; Cucciniello, R.; Paone, E.; Mauriello, F.; Tabanelli, T. A Short Overview on the Hydrogen Production Via Aqueous Phase Reforming (APR) of Cellulose, C6-C5 Sugars and Polyols. *Catalysts* **2019**, *9*, 917. [CrossRef]
14. Deng, W.; Zhang, Q.; Wang, Y. Catalytic transformations of cellulose and cellulose-derived carbohydrates into organic acids. *Catal. Today* **2014**, *234*, 31–41. [CrossRef]
15. Signoretto, M.; Taghavi, S.; Ghedini, E.; Menegazzo, F. Catalytic Production of Levulinic Acid (LA) from Actual Biomass. *Molecules* **2019**, *24*, 2760. [CrossRef] [PubMed]
16. Sudarsanam, P.; Zhong, R.; Bosch, S.V.D.; Coman, S.M.; Parvulescu, V.I.; Sels, B.F. Functionalised heterogeneous catalysts for sustainable biomass valorisation. *Chem. Soc. Rev.* **2018**, *47*, 8349–8402. [CrossRef]
17. Sarkar, J.; Bhattacharyya, S. Application of Graphene and Graphene-Based Materials in Clean Energy-Related Devices Minghui. *Arch. Thermodyn.* **2012**, *33*, 23–40. [CrossRef]
18. Lam, E.; Luong, J.H. Carbon Materials as Catalyst Supports and Catalysts in the Transformation of Biomass to Fuels and Chemicals. *ACS Catal.* **2014**, *4*, 3393–3410. [CrossRef]
19. Taghavi, S.; Norouzi, O.; Tavasoli, A.; di Maria, F.; Signoretto, M.; Menegazzo, F.; di Michele, A. Catalytic conversion of Venice lagoon brown marine algae for producing hydrogen-rich gas and valuable biochemical using algal biochar and Ni/SBA-15 catalyst. *Int. J. Hydrogen Energy* **2018**, *43*, 19918–19929. [CrossRef]
20. Xu, Q.; Zhu, Z.; Tian, Y.; Deng, J.; Shi, J.; Fu, Y. Sn-MCM-41 as Efficient Catalyst for the Conversion of Glucose into 5-Hydroxymethylfurfural in Ionic Liquids. *BioResources* **2013**, *9*, 303–315. [CrossRef]
21. Ya'Aini, N.; Amin, N.A.S.; Endud, S. Characterization and performance of hybrid catalysts for levulinic acid production from glucose. *Microporous Mesoporous Mater.* **2013**, *171*, 14–23. [CrossRef]

22. Ya'Aini, N.; Amin, N.A.S.; Asmadi, M. Optimization of levulinic acid from lignocellulosic biomass using a new hybrid catalyst. *Bioresour. Technol.* **2012**, *116*, 58–65. [CrossRef]
23. Cao, X.; Peng, X.; Sun, S.; Zhong, L.; Chen, W.; Wang, S.; Sun, S. Hydrothermal conversion of xylose, glucose, and cellulose under the catalysis of transition metal sulfates. *Carbohydr. Polym.* **2015**, *118*, 44–51. [CrossRef] [PubMed]
24. Ghedini, E.; Menegazzo, F.; Signoretto, M.; Manzoli, M.; Pinna, F.; Strukul, G. Mesoporous silica as supports for Pd-catalyzed H2O2 direct synthesis: Effect of the textural properties of the support on the activity and selectivity. *J. Catal.* **2010**, *273*, 266–273. [CrossRef]
25. Brunauer, S.; Emmett, P.H. The Use of Low Temperature van der Waals Adsorption Isotherms in Determining the Surface Areas of Various Adsorbents. *J. Am. Chem. Soc.* **1937**, *59*, 2682–2689. [CrossRef]
26. Barrett, E.P.; Joyner, L.G.; Halenda, P.P. The Determination of Pore Volume and Area Distributions in Porous Substances. I. Computations from Nitrogen Isotherms. *J. Am. Chem. Soc.* **1951**, *73*, 373–380. [CrossRef]
27. Gregg, S.J.; Sing, K.S.W. *Adsorption, Surface Area end Porosity*; Academic Press: New York, NY, USA, 1967.
28. Union, I.; Pure, O.F.; Chemistry, A. International union of pure commission on colloid and syrface chemistry including catalysis reporting physisorption data for gas/solid systems with Special Reference to the Determination of Surface Area and Porosity. *Area* **1985**, *57*, 603–619.
29. Ungár, T. Microstructural parameters from X-ray diffraction peak broadening. *Scr. Mater.* **2004**, *51*, 777–781. [CrossRef]
30. Hadjiev, V.G.; Iliev, M.; Vergilov, I.V. The Raman spectra of Co3O4. *J. Phys. C Solid State Phys.* **1988**, *21*, L199–L201. [CrossRef]
31. Rivas-Murias, B.; Salgueiriño, V. Thermodynamic CoO-Co3O4 crossover using Raman spectroscopy in magnetic octahedron-shaped nanocrystals. *J. Raman Spectrosc.* **2017**, *14*, 640–841. [CrossRef]
32. Gouadec, G.; Colomban, P. Raman Spectroscopy of nanomaterials: How spectra relate to disorder, particle size and mechanical properties. *Prog. Cryst. Growth Charact. Mater.* **2007**, *53*, 1–56. [CrossRef]
33. Sharma, R.K.; Reddy, G.B. Synthesis, and characterization of α-MoO3microspheres packed with nanoflakes. *J. Phys. D Appl. Phys.* **2014**, *47*, 65305. [CrossRef]
34. Cherian, C.T.; Reddy, M.V.; Haur, S.C.; Chowdari, B.V.R. Interconnected Network of CoMoO4 Submicrometer Particles As High Capacity Anode Material for Lithium Ion Batteries. *ACS Appl. Mater. Interfaces* **2013**, *5*, 918–923. [CrossRef] [PubMed]
35. Julien, C.; Massot, M.; Poinsignon, C. Lattice vibrations of manganese oxides. *Spectrochim. Acta Part A Mol. Biomol. Spectrosc.* **2004**, *60*, 689–700. [CrossRef]
36. Lorite, I.; Romero, J.J.; Fernández, J.F. Effects of the agglomeration state on the Raman properties of Co3O4 nanoparticles. *J. Raman Spectrosc.* **2012**, *43*, 1443–1448. [CrossRef]
37. Klissurski, D.G.; Uzunova, E. Cation-deficient nano-dimensional particle size cobalt–manganese spinel mixed oxides. *Appl. Surf. Sci.* **2003**, *214*, 370–374. [CrossRef]
38. Kovanda, F.; Rojka, T.; Dobešová, J.; Machovič, V.; Bezdička, P.; Obalová, L.; Jirátová, K.; Grygar, T.M. Mixed oxides obtained from Co and Mn containing layered double hydroxides: Preparation, characterization, and catalytic properties. *J. Solid State Chem.* **2006**, *179*, 812–823. [CrossRef]
39. Liu, B.; Ouyang, B.; Zhang, Y.; Lv, K.; Li, Q.; Ding, Y.; Li, J. Effects of mesoporous structure and Pt promoter on the activity of Co-based catalysts in low-temperature CO_2 hydrogenation for higher alcohol synthesis. *J. Catal.* **2018**, *366*, 91–97. [CrossRef]
40. Griboval, A.; Butel, A.; Ordomsky, V.V.; Chernavskii, P.A.; Khodakov, A.Y. Cobalt and iron species in alumina supported bimetallic catalysts for Fischer–Tropsch reaction. *Appl. Catal. A Gen.* **2014**, *481*, 116–126. [CrossRef]
41. Tavasoli, A.; Trépanier, M.; Abbaslou, R.M.M.; Dalai, A.K.; Abatzoglou, N. Fischer–Tropsch synthesis on mono- and bimetallic Co and Fe catalysts supported on carbon nanotubes. *Fuel Process. Technol.* **2009**, *90*, 1486–1494. [CrossRef]
42. Mai, K.; Elder, T.; Groom, L.H.; Spivey, J.J. Fe-based Fischer Tropsch synthesis of biomass-derived syngas: Effect of synthesis method. *Catal. Commun.* **2015**, *65*, 76–80. [CrossRef]
43. Kukushkin, R.; Bulavchenko, O.; Kaichev, V.; Yakovlev, V. Influence of Mo on catalytic activity of Ni-based catalysts in hydrodeoxygenation of esters. *Appl. Catal. B Environ.* **2015**, *163*, 531–538. [CrossRef]

44. Saghafi, M.; Heshmati-Manesh, S.; Ataie, A.; Khodadadi, A.A. Synthesis of nanocrystalline molybdenum by hydrogen reduction of mechanically activated MoO3. *Int. J. Refract. Met. Hard Mater.* **2012**, *30*, 128–132. [CrossRef]
45. Rodríguez, J.A.; Chaturvedi, S.; Hanson, J.C.; Brito, J.L. Reaction of H2and H2S with CoMoO4and NiMoO4: TPR, XANES, Time-Resolved XRD, and Molecular-Orbital Studies. *J. Phys. Chem. B* **1999**, *103*, 770–781. [CrossRef]
46. Khodakov, A.Y.; Chu, W.; Fongarland, P. Advances in the Development of Novel Cobalt Fischer–Tropsch Catalysts for Synthesis of Long-Chain Hydrocarbons and Clean Fuels. *Chem. Rev.* **2007**, *107*, 1692–1744. [CrossRef] [PubMed]
47. Bragança, L.; Ojeda, M.; Fierro, J.L.G.; da Silva, M.P. Bimetallic Co-Fe nanocrystals deposited on SBA-15 and HMS mesoporous silicas as catalysts for Fischer–Tropsch synthesis. *Appl. Catal. A Gen.* **2012**, *423*, 146–153. [CrossRef]
48. Srivastava, S.; Jadeja, G.C.; Parikh, J.K. A versatile bi-metallic copper–cobalt catalyst for liquid phase hydrogenation of furfural to 2-methylfuran. *RSC Adv.* **2016**, *6*, 1649–1658. [CrossRef]
49. Peng, Y.; Chang, H.; Dai, Y.; Li, J. Structural and Surface Effect of MnO2 for Low Temperature Selective Catalytic Reduction of NO with NH3. *Procedia Environ. Sci.* **2013**, *18*, 384–390. [CrossRef]
50. Mosallanejad, S.; Dlugogorski, B.Z.; Kennedy, E.; Stockenhuber, M. On the Chemistry of Iron Oxide Supported on γ-Alumina and Silica Catalysts. *ACS Omega* **2018**, *3*, 5362–5374. [CrossRef]
51. Liu, C.; Zhang, Z.; Zhai, X.; Wang, X.; Gui, J.Z.; Zhang, C.; Zhu, Y.; Li, Y. Synergistic effect between copper and different metal oxides in the selective hydrogenolysis of glucose. *New J. Chem.* **2019**, *43*, 3733–3742. [CrossRef]
52. Watanabe, M.; Bayer, F.; Kruse, A. Oil formation from glucose with formic acid and cobalt catalyst in hot-compressed water. *Carbohydr. Res.* **2006**, *341*, 2891–2900. [CrossRef]
53. Aman, D.; Radwan, D.; Ebaid, M.; Mikhail, S.; van Steen, E. Comparing nickel and cobalt perovskites for steam reforming of glycerol. *Mol. Catal.* **2018**, *452*, 60–67. [CrossRef]
54. Cheng, Z.; Everhart, J.L.; Tsilomelekis, G.; Nikolakis, V.; Saha, B.; Vlachos, D.G.; Vlachos, D. Structural analysis of humins formed in the Brønsted acid catalyzed dehydration of fructose. *Green Chem.* **2018**, *20*, 997–1006. [CrossRef]
55. Swift, T.D.; Nguyen, H.; Anderko, A.; Nikolakis, V.; Vlachos, D.G. Tandem Lewis/Brønsted homogeneous acid catalysis: Conversion of glucose to 5-hydoxymethylfurfural in an aqueous chromium(iii) chloride and hydrochloric acid solution. *Green Chem.* **2015**, *17*, 4725–4735. [CrossRef]
56. Kuninobu, Y.; Uesugi, T.; Kawata, A.; Takai, K. Manganese-Catalyzed Cleavage of a Carbon-Carbon Single Bond between Carbonyl Carbon and α-Carbon Atoms of Ketones. *Angew. Chem. Int. Ed.* **2011**, *50*, 10406–10408. [CrossRef]

© 2020 by the authors. Licensee MDPI, Basel, Switzerland. This article is an open access article distributed under the terms and conditions of the Creative Commons Attribution (CC BY) license (http://creativecommons.org/licenses/by/4.0/).

Article

Pd/Au Based Catalyst Immobilization in Polymeric Nanofibrous Membranes via Electrospinning for the Selective Oxidation of 5-Hydroxymethylfurfural

Danilo Bonincontro [1], Francesco Fraschetti [1], Claire Squarzoni [1], Laura Mazzocchetti [1,2], Emanuele Maccaferri [1], Loris Giorgini [1,2], Andrea Zucchelli [2,3], Chiara Gualandi [2,4], Maria Letizia Focarete [4] and Stefania Albonetti [1,*]

[1] Department of Industrial Chemistry Toso Montanari, University of Bologna, Viale Risorgimento, 4, 40136 Bologna, Italy; danilo.bonincontro2@unibo.it (D.B.); fr.fraschetti@yahoo.it or francesco.fraschetti@studio.unibo.it (F.F.); claire.squarzoni@gmail.com or claire.squarzoni@etu.chimieparistech.psl.eu (C.S.); laura.mazzocchetti@unibo.it (L.M.); emanuele.maccaferri3@unibo.it (E.M.); loris.giorgini@unibo.it (L.G.)
[2] Interdepartmental Center for Industrial Research on Advanced Applications in Mechanical Engineering and Materials Technology, CIRI-MAM, University of Bologna, Viale Risorgimento, 2, 40136 Bologna, Italy; a.zucchelli@unibo.it (A.Z.); c.gualandi@unibo.it (C.G.)
[3] Department of Industrial Engineering, University of Bologna, Viale Risorgimento, 2, 40136 Bologna, Italy
[4] Department of Chemistry Giacomo Ciamician and INSTM UdR of Bologna, University of Bologna, Via Selmi, 2, 40126 Bologna, Italy; marialetizia.focarete@unibo.it
* Correspondence: stefania.albonetti@unibo.it

Received: 6 December 2019; Accepted: 30 December 2019; Published: 1 January 2020

Abstract: Innovative nanofibrous membranes based on Pd/Au catalysts immobilized via electrospinning onto different polymers were engineered and tested in the selective oxidation of 5-(hydroxymethyl)furfural in an aqueous phase. The type of polymer and the method used to insert the active phases in the membrane were demonstrated to have a significant effect on catalytic performance. The hydrophilicity and the glass transition temperature of the polymeric component are key factors for producing active and selective materials. Nylon-based membranes loaded with unsupported metal nanoparticles were demonstrated to be more efficient than polyacrylonitrile-based membranes, displaying good stability and leading to high yield in 2,5-furandicarboxylic acid. These results underline the promising potential of large-scale applications of electrospinning for the preparation of catalytic nanofibrous membranes to be used in processes for the conversion of renewable molecules.

Keywords: polymeric catalytic membranes; electrospinning; HMF oxidation

1. Introduction

Monomer and polymer production from renewable feedstocks has become a relevant research target with the aim of providing more environmental friendly solutions to the actual fossil-based market [1–6]. In this framework, 5-(hydroxymethyl)furfural (HMF) is recognized as an ideal platform molecule to develop different green products, since it can be obtained via acid-catalyzed dehydration of biomass-derived sugars [7–9], and, in turn, it can be converted into a wide range of different high added value chemicals [10–13]. Among the HMF products, 2,5-furandicarboxylic acid (FDCA, Scheme 1) has been identified as one of the most interesting [14], since it can be considered as the bioderived counterpart of terephthalic acid for the production of polyesters [15], such as polyethylene 2,5-furandicarboxylate

(PEF), being the latter the potential candidate to replace polyethylene terephthalate (PET) in bottle production [16,17].

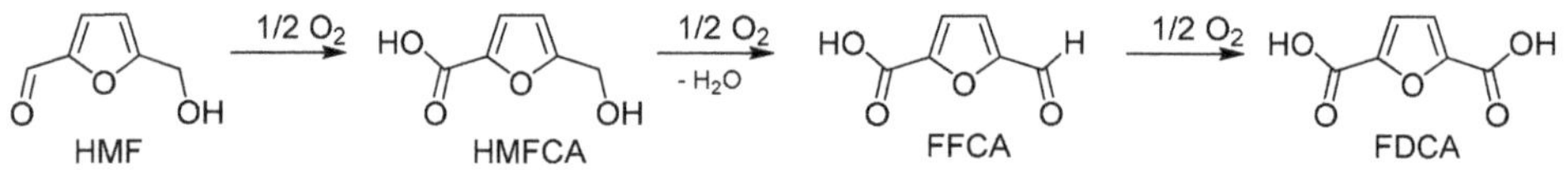

Scheme 1. 5-(hydroxymethyl)furfural selective oxidation to 2,5-furandicarboxylic acid.

First attempts to industrially convert HMF to FDCA relied on the technologies developed for terephthalic acid production [18]. However, the use of corrosive solvent, homogeneous catalysts and harsh operative conditions forced the investigation of other catalytic systems in order to overcome such constraints. In this regard, precious metal nanoparticles showed their high potential, Au-based systems being the most investigated [19–21]. In the definition of the properties of such catalytic systems, several parameters seem to be fundamental such as: nanoparticle composition and dimension, support/nanoparticle interaction and support surface textural properties. The fine tuning of such parameters can lead to very active systems. For instance, by alloying Au with another metal (such as Cu [22,23] or Pd [24,25]) it is possible to effectively enhance FDCA yield, or by choosing suitable supports catalytic activity can be positively affected [26] and/or base addition can be avoided [27–29]. Besides the textural properties that have been shown to play a prominent role, pore dimension and acid/base sites can be considered of crucial importance [30]. Thus, it is evident that the design and synthesis of materials that possess suitable features can lead to optimal catalytic activity.

As far as the setups employed to carry out this reaction, batch ones are the most studied. However, the economical sustainability of this kind of approach is still a concern [31]. Thus, to address this issue, efforts must be devoted to process intensification, for instance by evaluating the possibility of developing inexpensive catalytic systems to perform this reaction continuously [32,33], and, in this frame, recent reports have highlighted the interest of both industrial [34,35] and academic research [36,37]. In this context, the use of catalytic membranes is known to provide several advantages, making them efficient tools for applications in several industrial fields [38]. In the case of HMF oxidation, considering that the reaction is carried out under mild operative conditions (70–120 °C), it is possible that the use of composite polymeric membranes might be advantageous. Along with all the advantages related to the more traditional inorganic membranes, these materials are characterized by low production costs, ease of handling and tunability of their properties [39].

Electrospinning provides a convenient approach for the preparation and scale-up of membranes made of continuous sub-micrometric fibers characterized by large surface area and porosity [40,41]. This could represent an interesting strategy for the production of catalytic membranes, which can be used in processes for biomass valorization. Briefly, this technology uses electrostatic forces to uniaxially stretch a viscoelastic jet derived from a polymer solution to produce fibers having diameters ranging from a few tenths of nanometers to a few micrometers, collected as nonwovens with mesh porosity typically higher than 80% and pore diameters that can vary from a few to tens of micrometers.

Electrospun membranes are currently being investigated as materials for the support of heterogeneous catalysts by following different technological approaches, the main one being the production of ceramic fibers that support metal nanoparticles [40]. In this case, a polymer solution containing ceramic precursors and metal salts is electrospun and subsequently heat treated under inert gas to eliminate the organic components and reduce the metal precursor to metal nanoparticles [42–45]. This approach permits the achievement of high catalytic performances by exploiting the high surface area of the fibers but suffers from the high fragility of the completely inorganic nonwoven. Conversely, by keeping unaltered the organic polymeric component, membrane flexibility and handling can be massively improved. Following this approach, polymeric electrospun nanofibers have been decorated at the surface with metal nanoparticles (NPs) [46,47] in an elegant and effective way that exploits polymer bulk properties and maximizes the catalytic effect. However, NP immobilization at a fiber surface requires a further step in the production process that is time consuming and hardly applicable at the industrial level. Moreover, leaching of metal NPs from a fiber surface cannot be excluded.

In this work, electrospinning is used in a simple and scalable single-step approach for the production of electrospun polymer-inorganic catalytic membranes, potentially suitable for batch and continuous processes. Membranes were manufactured by incorporating preformed Au and Au/Pd NPs (Au/Pd molar ratio 6, which was demonstrated to promote the highest catalytic activity in this reaction [24]) and TiO_2 in the starting polymeric solutions. To optimize catalytic activity and stability, two different polymers have been tested—i.e., polyacrylonitrile (PAN) and Nylon 6,6 (NYL)—loaded with Au and alloyed Au/Pd NPs, either directly supported on TiO_2 or simply combined with TiO_2 during electrospinning. The catalytic activity of catalysts contained in different electrospun membranes towards HMF oxidation to FDCA has been investigated with the goal of highlighting the effect of polymer/inorganic combination on membrane performance. The materials were evaluated in batch experiments to assess the viability for use of electrospun polymer-based catalytic membranes in the conversion of renewable molecules in water.

2. Materials and Methods

2.1. Materials

Polyacrylonitrile (PAN, M_w = 1.5 × 10^5 g/mol) was purchased from Sigma-Aldrich (St. Louis, MO, USA). Nylon 6,6 (NYL, Zytel® E53 NC010) was kindly provided by DuPont (Wilmington, DE, USA). Dimethylformamide (DMF), formic acid (FA) and chloroform (CLF), $HAuCl_4$, $PdCl_2$, glucose, NaOH and polyvinylpyrrolidone (PVP) were purchased from Sigma Aldrich and were used without further purification. HMF (purity > 99%) was purchased from AVABiochem (Muttenz, Switzerland) and used without any purification.

2.2. Nanoparticle Synthesis

Au-based nanoparticle synthesis was performed as previously reported [24]. As a general approach, a suitable amount of the metal precursors ($HAuCl_4$ and $PdCl_2$) were dissolved in water. Once dissolved, glucose, NaOH and the stabilizing agent (PVP) were added to the solution and allowed to react for 2.5 min at 95 °C under solvent reflux. Then, the resulting nanoparticle suspension was concentrated using 50 kDa Amicon Ultra filters (Millipore, Burlington, MO, USA) to eliminate excess water. The concentrated suspension was impregnated onto TiO_2 in order to achieve a metal loading (pristine Au or Au + Pd with 6/1 metal ratio) of 1.5 wt.%. After the impregnation, solvent was evaporated by thermal treatment at 120 °C. Alternatively, another batch of the Au_6Pd_1 NPs colloidal suspension was washed with formic acid in two subsequent filtrations to replace water.

2.3. Production of Electrospun Membranes

Electrospun non-woven mats with random arrangement of fibers were fabricated using an electrospinning machine (Spinbow s.r.l, Bologna, Italy). Briefly, the electrospinning apparatus was composed of a high voltage power supply, a syringe pump and a glass syringe containing the polymer solution connected to a stainless-steel blunt-ended needle through a polytetrafluoroethylene tube. A grounded plate collector was vertically positioned below the tip of the needle. Electrospinning was performed at room temperature (RT) and relative humidity 40%–50%.

PAN and nylon based membranes were prepared using a stainless-steel blunt needle (inner diameter = 0.84 mm) and collected on a plate collector (25 × 25 cm^2). First membranes containing plain TiO_2 and TiO_2 supported Au_6Pd_1 nanoparticles, together with a reference polymeric membrane, were prepared. Then, membranes containing unsupported Au_6Pd_1 nanoparticles and TiO_2 independently added and just Au_6Pd_1 nanoparticles were also produced. The residual solvent in the membranes was removed by thermal treatment at 80 °C for 3 h in static air. The mass loss due to the drying was around 1%–3% for all the membranes.

The detailed description of solution preparation and electrospinning conditions for all the produced samples are reported in the Supporting Information section (Tables S1–S3, Figures S1 and S2).

2.4. Characterization Methods

X-ray diffraction (XRD) measurements were carried out at room temperature with a Bragg/Brentano diffractometer (X'pertPro PANalytical - Malvern Panalytical Ltd, Malvern, UK) equipped with a fast X'Celerator detector, using a Cu anode as the X-ray source (K_α = 1.5418 Å). For all samples, diffractograms were recorded in the range 35–44° 2θ, counting for 1000 s every 0.1° 2θ step. Crystallite size values were calculated using the Scherrer equation from the full width at half maximum intensity measurements.

Scanning electron microscopy (SEM) observations were carried out by using a Leica Cambridge Stereoscan 360 scanning electron microscope (Leica, Cambridge, UK) at an accelerating voltage of 20 kV, on samples sputter-coated with gold. The distribution of fiber diameters was determined through the measurement of about 150 fibers and the results were given as the average diameter ± standard deviation (SD).

Transmission electron microscopy (TEM) observations were carried out by using a FEI Tecnai F20 microscope (Thermo Fisher Scientific, Waltham, MA, USA) equipped with a Schottky emitter and operating at 200 KeV. The fibers were electrospun directly on a TEM copper grid (100 mesh).

Differential scanning calorimetry (DSC) measurements were carried out using a TA Instruments Q100 DSC (Thermal Analysis Instruments, New Castle, PA, USA) equipped with the liquid nitrogen cooling system accessory. DSC scans of electrospun membranes were performed in helium atmosphere at a heating rate of 20 °C/min. The glass transition temperature (T_g) was taken at half-height of the glass transition heat capacity step, while the melting temperature (T_m) was taken at the peak maximum of the melting endotherm.

Thermogravimetric analyses (TGA) were carried out using a TGA Q500 thermogravimetric analyzer (TA Instruments, New Castle, DE, USA). Analyses were performed from RT to 700 °C, at a heating rate of 10 °C/min, under air flow.

The BET specific surface area of each catalyst was determined by N_2 absorption–desorption at liquid N_2 temperature, using a Sorpty 1750 Fison instrument (Micromeritics Instruments Corporation, Norcross, GA, USA). Prior N_2 absorption samples were outgassed at 50 °C.

2.5. Catalytic Tests

A lab scale autoclave reactor (100 mL capacity from Parr) (Parr Instrument Company, Moline, IL 61265-1770, USA), equipped with a mechanical stirrer (0–600 rpm) and sensors for temperature and pressure measurement, was used to carry out the catalytic tests. An aqueous solution of HMF (25 mL, 18 mM) and NaOH (NaOH/HMF molar ratio = 2) and the membrane previously cut into small squared pieces of 1 cm^2 (HMF/total metal molar ratio = 100) were loaded in the reactor. To perform the reactivity tests, the autoclave reactor was purged three times with O_2 (2 bar) and then pressurized at 10 bar. The temperature was increased to the set point and the reaction mixture was stirred at 600 rpm for the whole duration of the experiment (4 h if not stated differently). As reaction initial time (time zero) for the reaction was considered when the set point temperature was reached (after 10 min). At the end of the reaction, the reactor was cooled down to RT and the solution was filtered. Then, the reaction mixture was diluted five times and then analyzed with an Agilent Infinity 1260 liquid chromatograph equipped with an Aminex HPX-87H 300 mm × 7.8 mm column (0.005 M H_2SO_4 solution as mobile phase) and a DAD. Compound quantifications were performed from the peak areas after calibration using commercially available samples. Recycling tests have been performed by reusing the recovered membrane without any further cleaning procedure.

3. Results and Discussion

The preparation of monometallic Au and bimetallic Au/Pd (molar ration 6/1) was optimized elsewhere [24] and led to the formation of small nanoparticles, having mean nanoparticle diameter of 4 and 4.5 nm for Au and AuPd NPs, respectively (revealed by XRD analysis and TEM—Table S4). Their impregnation on TiO_2, with a metal lading of 1.5%, leads to slight mean particle size increase (Figures S3 and S4).

3.1. PAN-Based Catalytic Membranes

3.1.1. Characterization

The different types of NPs supported onto TiO_2 were incorporated into PAN electrospun membranes, as described in detail in the Supporting Information, and the resulting morphology was observed by SEM (Figure 1) and TEM (Figure 2).

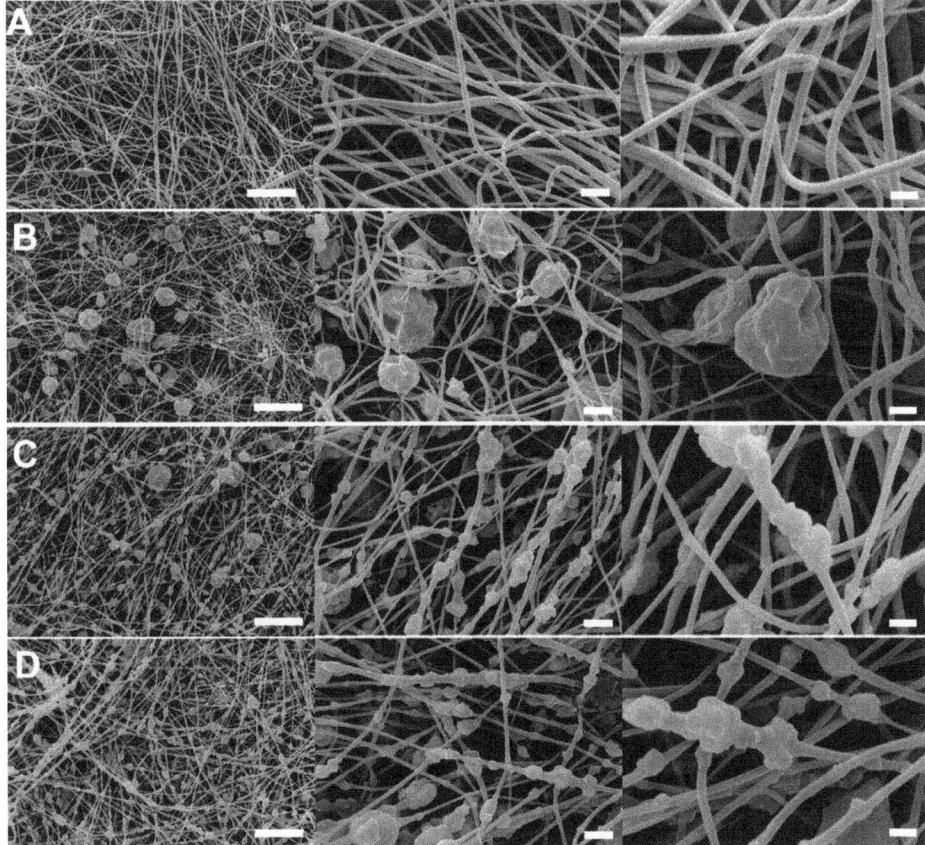

Figure 1. SEM images acquired at different magnifications of: (**A**) PAN; (**B**) PAN+TiO_2; (**C**) PAN + Au/TiO_2 and (**D**) PAN + AuPd/TiO_2. Scale bars: 10 µm (first column); 2 µm (second column) and 1 µm (third column).

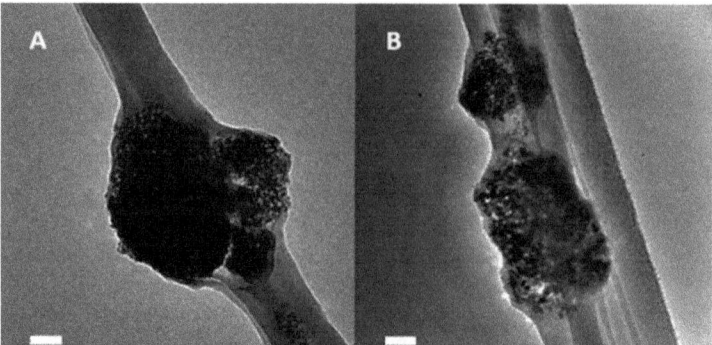

Figure 2. TEM analysis of PAN-Au/TiO$_2$ and PAN-AuPd/TiO$_2$ samples (**A**,**B**, respectively). Scale bars: 100 nm.

For all types of PAN-based membranes the continuous fibers have diameters in the range 200–300 nm, similarly, to unloaded PAN membrane (Figure 1A). In PAN + TiO$_2$ (Figure 1B) the TiO$_2$ NPs are round-shaped, with a rough and corrugated surface and with a broad distribution of dimensions. The smallest particles (in the range of 0.5–1 μm) are clearly incorporated along the fibers while the biggest ones (up to 5 μm in diameter) are entrapped between the pores generated by the non-woven structure. In both PAN + Au/TiO$_2$ and PAN + AuPd/TiO$_2$ samples, the agglomerates are randomly distributed along fiber axis, mostly confined inside the PAN fibers. The particles are small (diameter 0.1–1 μm) though being bigger than fiber diameter, thus generating a "pearl necklace" morphology. The fact that large aggregates are entrapped in the pore of PAN + TiO$_2$ while small particles are mostly incorporated in the fibers of PAN + Au/TiO$_2$ and PAN + AuPd/TiO$_2$ can be ascribable to the original different particle dimensions (Table S4).

TEM inspection (Figure 2) indicates the successful incorporation of Au/TiO$_2$ and AuPd/TiO$_2$ in PAN fibers and reveals that the particles observed in SEM images are aggregates of TiO$_2$.

TGA analysis (Figure S5) confirms the organic/inorganic composition of electrospun membranes. PAN membrane shows a constant weight up to 250 °C (apart from a modest weight loss of 1% at low temperature ascribable to residual DMF evaporation, see inset), followed by a sharp weight loss of about 35% and a slower second weight loss starting from about 500 °C, with a negligible weight residue at 700 °C, in line with previous results [48,49]. PAN membranes loaded with supported NPs show the same degradation pattern, with weight losses proportional to the PAN content, while the residual weight at 700 °C corresponds to the inorganic phase not subjected to thermal degradation and was in the range 61%–62% for all membranes, as a proof of process reproducibility.

3.1.2. Catalytic Tests

Membranes were tested in the liquid phase oxidation of HMF. This molecule has two groups that can be oxidized: the alcoholic and the aldehydic group. The complete or partial oxidation of one or both groups may lead to the formation of different products (Scheme 1). The reaction on Au-based catalysts has been generally described in two steps: (i) the oxidation of aldehydic group to 5-hydroxymethyl-2-furancarboxylic acid (HMFCA) and (ii) the oxidation of alcoholic group—through the formation of 5-formyl-2-furancarboxylic acid (FFCA)—to 2,5-furandicarboxylic acid (FDCA). 2,5-diformylfuran (DFF) was not generally observed in the course of the reaction with gold-base catalysts. Our study (Table 1) indicated that the plain supporting phases (either PAN or PAN + TiO$_2$) were inactive in the oxidation (entry 1 and 2), while forming very small amounts of HMFCA and by-products derived from HMF degradation favored by the high pH, in agreement with previous studies [22]. On the other hand, when Au-decorated TiO$_2$ was inserted in the membrane network, the resulting materials display a certain activity (entry 3), which was far lower if compared to the powder catalyst (entry 5). This could be attributed to the fewer active sites exposed in the membrane respect to the overall active sites of the powder. The use of the bimetallic system induced a significant increase of the catalytic performance of the membrane: HMF conversion increased from 69% to 94%, and a small amount of FDCA (2%) was also detected with this catalyst (entry 4). The improved performances of the bimetallic system compared to the Au monometallic system is correlated to the cooperative effect of the two metals in the alloyed system, as demonstrated in previous papers [24,50]. However, the catalytic activity of the AuPd-containing membrane was far lower than the one of the respective powder sample (entry 4 and 6, respectively), indicating that, also in this case, substrate access to the catalyst active sites is hindered by the polymer. It is indeed worth to point out that catalytic tests are carried out at 70 °C, a temperature significantly lower than the polymer glass transition (T_g = 108 °C), and the polymeric phase might thus represent a diffusion barrier which is even harder to overcome in the glassy state. In order to overcome this problem, the PAN + AuPd/TiO$_2$ membrane was treated at a temperature higher than the PAN glass transition temperature. This has been done with the aim to allow polymer chain mobility, which could lead, in turn, to an improved exposure of the active sites. Alternatively, another batch of untreated membrane, was calcined at 300 °C, with the aim to understand if the thermal induced PAN cyclisation has a positive effect on the membrane catalytic performances. Nevertheless, these tests, also performed at higher temperatures (Figures S6 and S7) were not successful in recovering a higher catalytic activity.

Table 1. Catalytic performances of PAN-derived materials. Reaction conditions: 4 h, 70 °C, O$_2$ pressure 10 bar, 25 mL water, HMF (0.018 M), HMF:NaOH molar ratio 1:2 and HMF:metal molar ratio 100:1.

Entry	Sample	HMF Conversion (%)	HMFCA Yield (%)	FFCA Yield (%)	FDCA Yield (%)	C-LOSS (%)
1	PAN	39	3	1	0	35
2	PAN+TiO$_2$	64	4	0	0	60
3	PAN + Au/TiO$_2$	69	21	5	0	43
4	PAN + AuPd/TiO$_2$	94	65	24	2	3
5	Au/TiO$_2$	100	85	10	5	0
6	AuPd/TiO$_2$	100	30	10	60	0

3.2. Nylon-Based Membranes

With the aim of solving the diffusion problems observed with PAN nanofibers, Nylon 6,6 (NYL) was thus selected as matrix polymer for catalytic membrane production. Indeed, Nylon 6,6 has a lower glass transition (T_g = 50–60 °C) compared to PAN coupled with a high melting temperature (Figure S8). These properties should allow good temperature stability and a slightly higher hydrophilicity with respect to PAN fibers. Stemming from previously obtained results that highlighted better performances for bimetallic systems, nanofibers were produced solely with TiO$_2$ supported AuPd nanoparticles (NYL + AuPd/TiO$_2$). Together with the reference NPs free fibers (i.e., plain NYL and NYL + TiO$_2$), two additional catalytic membranes were also produced, one containing independently added TiO$_2$ and unsupported AuPd NPs (NYL + AuPd + TiO$_2$) and the other containing just the unsupported AuPd NPs (NYL + AuPd). The latter two samples aimed at exploring the ability of the nanofibrous polymeric systems to stabilize the metallic nanoparticles and the role of TiO$_2$ in the complex catalytic medium. While the procedure for adding titania supported AuPd nanoparticles (AuPd/TiO$_2$) was similar to the previously applied method for PAN fibers (with some limitation in particle content due to high-concentration suspension stability issues during the process), addition of unsupported bimetallic NPs was possible due to the solvent system used for nylon electrospinning (formic acid/chloroform) that proved to be slightly water tolerant. Formic acid (FA) was indeed used to replace water as much as possible as NPs solvent (17 wt.% water residual tolerated) with subsequent washing/centrifugation steps that did not alter NPs dimensions or promoted their aggregation (Figures S3 and S4). Such FA suspension was then used in a procedure similar to the one applied for plain nylon nanofibers or to the procedure for obtaining NYL + TiO$_2$ nanofiber to attain NYL + AuPd and NYL + AuPd + TiO$_2$ catalytic membranes respectively.

3.2.1. Characterization

SEM micrographs, recorded for all the obtained nanofibrous samples (Figure 3), show thin fibers with diameters ranging in 350–80 nm span (Table 2).

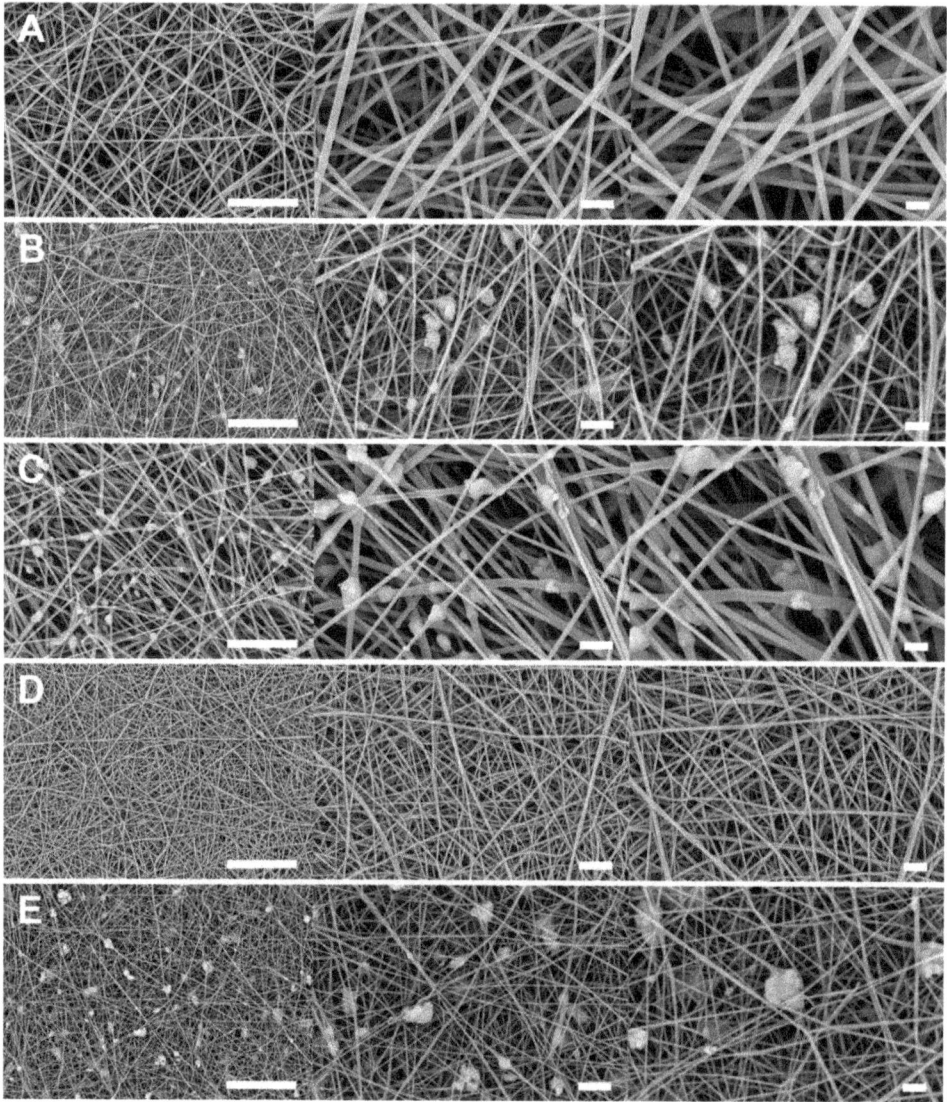

Figure 3. SEM images acquired at different magnifications of: (**A**) NYL; (**B**) NYL+TiO$_2$; (**C**) NYL + AuPd/TiO$_2$; (**D**) NYL + AuPd and (**E**) NYL + AuPd + TiO$_2$. Scale bars: 10 µm (first column); 2 µm (second column) and 1 µm (third column).

Table 2. NYL-based membranes: fiber average diameter and specific surface area of the electrospun samples.

Sample	Average Diameter (nm)	Specific Surface Area ($m^2\ g^{-1}$)
NYL	320 ± 70	<10
NYL + TiO_2	130 ± 20	36
NYL + Au/PdTiO_2	300 ± 80	32
NYL + AuPd	100 ± 20	<10
NYL + AuPd + TiO_2	80 ± 10	36

While the fibers are all thin and smooth, it was observed that inorganic component promoted fiber diameter thinning, and in particular, the presence of free unsupported AuPd NPs, decreased the average diameter around, or even below, 100 nm. Moreover, the addition of TiO_2 (both as plain titania or when NPs are supported on it) led to aggregates formation whose morphology well compared with the previously analyzed PAN based fibers. The dimension of such aggregates exceeded the fibers diameters, which by the way were way thinner than the pristine NYL counterpart, with some extroversion outside the smooth profile of the single filament. However, aggregates belonged to the fibers bulk, and were not simply leaning on the surface. The lower concentration of particles, due to the previously highlighted high-concentration suspension stability issues with the solvent system to be used for NYL fibers, provoked, in turn, minor aggregation phenomena with respect to the highly loaded PAN fibers, with no excess particle entrapment within the membrane pores. On the other side no aggregate was detected with just unsupported NPs, whose diameters were well below the fibers average size.

TGA measurements (Table S5) confirmed the trend observations with PAN fibers and the composition of the starting solution, meaning that no significant inorganic phase separation occurred during electrospinning, which might result in fibers with significant depletion of inorganic content in the fibers.

An additional important feature was also highlighted for nanofibrous membranes inorganic fillers such as titania, or NPs (either supported on titania or unsupported). Indeed, all of the "loaded fibers" are characterized by a higher T_g (about +10 °C) and a higher degree of crystallinity with respect to plain nylon nanofibers. The first observation accounts for a good dispersion of NP and TiO_2 supported NP within the polymer, so much that the mobility of the macromolecules was hindered by the interaction with the inorganic components, with a significant increase in glass transition. As far as the crystal phase is concerned, electrospinning is well renown to discourage crystal formation, while the presence of such fillers seems to help promoting crystallization during spinning, acting as nucleants [51]. This observation, while possibly not relevant for the mere catalytic activity, can be of paramount importance when considering the fibers mechanical properties, which are strongly influenced by their crystallinity, and is an important factor when dealing with membranes that should be able to withstand continuous flow conditions.

3.2.2. Catalytic Tests

The screening of the catalytic performances demonstrated the inability of the bare electrospun NYL membrane to catalyze the reaction in the studied conditions (Table 3, entry 1). Indeed, in this blank experiment, in the absence of the metallic active phase, more than 90% of the fed HMF was converted into degradation products and no oxidation occurred. The addition to the membrane of TiO_2-supported AuPd NPs made the membrane active in the formation of HMFCA and FFCA (Table 3, entry 2) but still low production of FDCA was evidenced.

Table 3. Catalytic performances of NYL-derived materials. Reaction conditions: 4 h, 70 °C, O_2 pressure 10 bar, 25 mL water, HMF (0.018 M), HMF:NaOH molar ratio 1:2 and HMF:metal molar ratio 100:1.

Entry	Sample	HMF Conversion (%)	HMFCA Yield (%)	FFCA Yield (%)	FDCA Yield (%)	C-LOSS
1	NYL	100	3	0	0	97
2	NYL + AuPd/TiO_2	100	51	28	6	15
3	NYL + AuPd	100	45	32	4	19
4	NYL + TiO_2 + AuPd	100	50	28	14	8

Contrarily to what was observed using PAN-based fibers, using NYL material, a strong effect of reaction temperature was observed on FDCA formation, with yield increasing from 6% up to 27% (Figure S9). Nylon, having lower glass transition temperature and higher hydrophilicity, seems to lead to lower diffusion issues.

To further increase the performance of the membrane, different materials were synthetized inserting pristine AuPd NPs colloids in the mixture to be electrospun. In particular, two different membranes have been prepared by (i) electrospinning of a suspension containing AuPd NPs colloids and nylon (NYL + AuPd sample) and (ii) electrospinning a suspension containing bare TiO_2, unsupported AuPd NPs and nylon (NYL + TiO_2 + AuPd sample). For both samples, after 4 h reaction time, HMF conversion was complete but considerable differences were seen among catalysts in term of product selectivity. In particular, the selectivity to FDCA increased significantly for TiO_2-containing sample. This feature could be ascribed to the effect of the increase of specific surface area obtained by introducing TiO_2 in the membrane (Table 3).

To further demonstrate the higher suitability of such synthetic protocol, two tests at higher reaction temperature (90 and 110 °C, Figure 4) were performed. These tests proved that temperature has a strong positive effect on FDCA yield, since it rose from 14% to 67% by increasing reaction temperature.

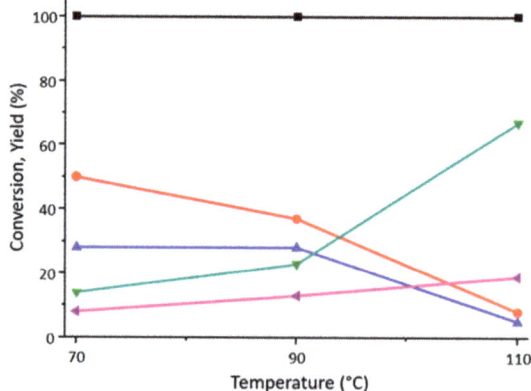

Figure 4. Reaction temperature effect on the catalytic performance of NYL + TiO_2 + AuPd sample. Operative conditions: 4 h, O_2 pressure 10 bar, 25 mL water, HMF concentration 0.018 M, HMF:NaOH molar ratio 1:2, HMF:(Au + Pd) molar ratio 100:1. Legend: ■ HMF Conversion, ● HMFCA yield, ▲ FFCA yield, ▼ FDCA yield, ◄ C-LOSS.

In order to evaluate the NYL + TiO_2 + AuPd membrane stability, reusability tests have been performed at 90 °C (Figure 5). Unexpectedly, NYL + TiO_2 + AuPd catalytic membrane showed a significant increase in activity after the first catalytic test, with FDCA yield rising from 19% to 34%.

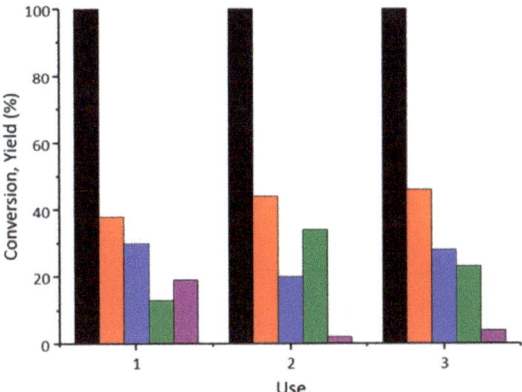

Figure 5. Reusability tests performed over NYL + TiO$_2$ + AuPd composite membranes. Operative conditions: 4 h, temperature 90 °C, O$_2$ pressure 10 bar, 25 mL water, HMF concentration 18 mM, HMF:NaOH molar ratio 1:2, HMF:(Au + Pd) molar ratio 100:1. Legend: ■ HMF Conversion, ■ HMFCA yield, ■ FFCA yield, ■ FDCA yield, ■ C-LOSS.

Since metal leaching during the reaction was excluded by chemical analysis (XRF analysis revealed that no Pd, Au of Ti species were dissolved in the reaction mixture), this initial activation effect could be attributed to a modification of the interaction between the active phase and the polymer during the reactivity experiment in water. Indeed, it is possible to hypothesize that catalytic active sites, which are hindered in the original membrane network, are made accessible by the use of the membrane in the reaction conditions, probably due to thermal induced movement of polymer chains during the reaction, which could occur because of the fact that membrane operates in temperature conditions (90 °C) higher than its glass transition temperature (50 °C). In this frame, SEM micrographs of fresh and materials used at 90 °C and 110 °C (Figure 6) strengthened this hypothesis. Indeed, while some fiber shrinking was observed in the pictures, no significant changes in fibers morphology were observed. Moreover, the inorganic content, as evaluated by TGA, stayed unaffected even after catalytic tests, confirming that no leaching of TiO$_2$ or metal nanoparticles occurred during the reaction.

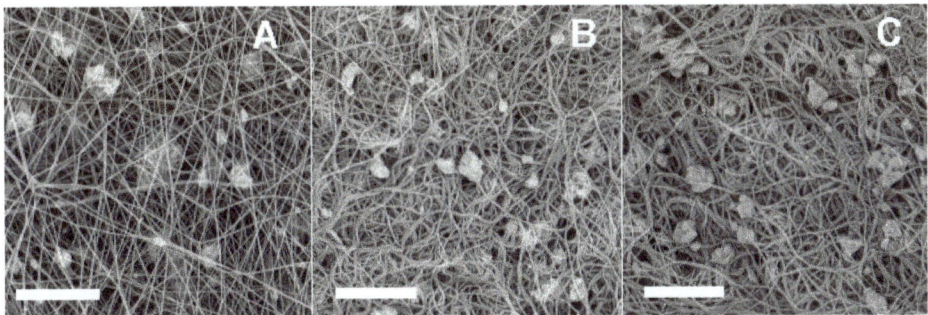

Figure 6. SEM micrographs of (**A**) NYL + AuPd + TiO$_2$; (**B**) NYL + AuPd + TiO$_2$ after reaction at 90 °C and (**C**) NYL + AuPd + TiO$_2$ after reaction at 110 °C; scale bar: 5 μm.

Preliminary analysis using the Attenuated Total Reflection coupled with Infrared Spectroscopy (ATR-FTIR) on fresh and used materials (Figure 7) showed that no furanic-compounds were present within both membranes, and no change in its chemical composition occurred during the reaction.

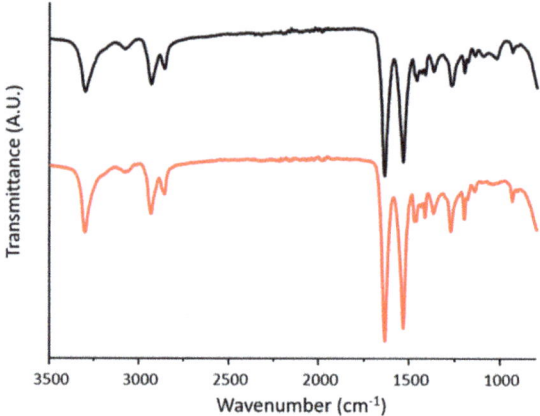

Figure 7. ATR analysis of NYL + TiO$_2$ + AuPd samples: fresh (black, up) and used (red, bottom).

4. Conclusions

This work demonstrated that AuPd-based nanosystems could be efficiently and quantitatively incorporated in a polymeric fibrous network via electrospinning. The resulting membranes possessed significant activity, which was strongly correlated to: (i) the chemical nature of the used polymeric matrix and (ii) the method of active-phase introduction. In particular, glass transition temperature lower than reaction temperature and high hydrophilicity positively affected membrane activity. As far as the introduction of the active phase is concerned, the addition of unsupported NPs and TiO$_2$ led to the production of the most active membrane. All electrospun materials showed high catalytic and structural stability and no leaching of active phase was observed during HMF oxidation in water. In conclusion, this work helped to gain fundamental knowledge on the features that play key roles in the utilization of heterogeneous catalysts supporting membranes for biomass oxidation in water, suggesting that current strategy could be a viable way for the future development of continuous processes based on this technology.

Supplementary Materials: The following are available online at http://www.mdpi.com/2227-9717/8/1/45/s1, Figure S1: DLS analyses of NPs suspensions; Figure S2: XRD analyses of dried NPs suspensions; Figure S3: Representative TEM images of Au/TiO$_2$ supported on TiO$_2$ and their size distribution histograms; Figure S4: Representative TEM images of AuPd/TiO$_2$ supported on TiO$_2$ and their size distribution histograms; Figure S5: TGA analyses of PAN-based membranes; Figure S6: Effect of thermal treatments on PAN + AuPd/TiO$_2$ membranes catalytic activity; Figure S7: Reaction temperature effect on the catalytic performance of PAN + AuPd/TiO$_2$ sample; Figure S8: DSC analyses of PAN and Nylon; Figure S9: Effect of reaction temperature on NYL + AuPd/TiO$_2$ membranes catalytic activity; Table S1: Amounts of polymer and inorganic fraction used to prepare the suspension in DMF and electrospinning processing conditions; Table S2: Amounts of polymer and inorganic fraction used to prepare FA/CHCl$_3$ suspensions; Table S3: Electrospinning processing conditions and average diameter obtained; Table S4: Characteristic of prepared catalysts and average diameters of metallic dimension estimated from XRD and TEM analysis; Table S5: Thermal characterization results for Nyl based nanofibers.

Author Contributions: D.B., L.M., E.M., and C.G. designed the different experiments and supported the interpretation of catalytic tests and material characterization; F.F. and C.S. synthesized the catalysts and carried out catalytic evaluation and characterization of materials; L.G., C.G., M.L.F., A.Z., and S.A. were involved in the writing and editing the manuscript. All authors have read and agreed to the published version of the manuscript.

Funding: This work was funded by SINCHEM Joint Doctorate Program-Erasmus Mundus Action (framework agreement N° 2013-0037).

Acknowledgments: The Italian Ministry of University and Research (MIUR) is acknowledged.

Conflicts of Interest: The authors declare no conflict of interest.

References

1. John, G.; Nagarajan, S.; Vemula, P.K.; Silverman, J.R.; Pillai, C.K.S. Natural monomers: A mine for functional and sustainable materials—Occurrence, chemical modification and polymerization. *Prog. Polym. Sci.* **2019**, *92*, 158–209. [CrossRef]
2. Ragauskas, A.J.; Williams, C.K.; Davison, B.H.; Britovsek, G.; Cairney, J.; Eckert, C.A.; Frederick, W.J.; Hallett, J.P.; Leak, D.J.; Liotta, C.L.; et al. The Path Forward for Biofuels and Biomaterials. *Science* **2006**, *311*, 484–489. [CrossRef] [PubMed]
3. Chen, G.-Q.; Patel, M.K. Plastics Derived from Biological Sources: Present and Future: A Technical and Environmental Review. *Chem. Rev.* **2012**, *112*, 2082–2099. [CrossRef] [PubMed]
4. Gandini, A. Monomers and Macromonomers from Renewable Resources. In *Biocatalysis in Polymer Chemistry*; John Wiley & Sons Ltd.: Hoboken, NJ, USA, 2010; pp. 1–33. ISBN 978-3-527-63253-4.
5. Delidovich, I.; Hausoul, P.J.C.; Deng, L.; Pfützenreuter, R.; Rose, M.; Palkovits, R. Alternative Monomers Based on Lignocellulose and Their Use for Polymer Production. *Chem. Rev.* **2016**, *116*, 1540–1599. [CrossRef] [PubMed]
6. Gandini, A.; Lacerda, T.M.; Carvalho, A.J.F.; Trovatti, E. Progress of Polymers from Renewable Resources: Furans, Vegetable Oils, and Polysaccharides. *Chem. Rev.* **2016**, *116*, 1637–1669. [CrossRef]
7. Portillo Perez, G.; Mukherjee, A.; Dumont, M.-J. Insights into HMF catalysis. *J. Ind. Eng. Chem.* **2019**, *70*, 1–34. [CrossRef]
8. Li, X.; Xu, R.; Yang, J.; Nie, S.; Liu, D.; Liu, Y.; Si, C. Production of 5-hydroxymethylfurfural and levulinic acid from lignocellulosic biomass and catalytic upgradation. *Ind. Crop. Prod.* **2019**, *130*, 184–197. [CrossRef]
9. Agarwal, B.; Kailasam, K.; Sangwan, R.S.; Elumalai, S. Traversing the history of solid catalysts for heterogeneous synthesis of 5-hydroxymethylfurfural from carbohydrate sugars: A review. *Renew. Sustain. Energy Rev.* **2018**, *82*, 2408–2425. [CrossRef]
10. Zhang, D.; Dumont, M.-J. Advances in polymer precursors and bio-based polymers synthesized from 5-hydroxymethylfurfural. *J. Polym. Sci. Part A Polym. Chem.* **2017**, *55*, 1478–1492. [CrossRef]
11. Hu, L.; Lin, L.; Wu, Z.; Zhou, S.; Liu, S. Recent advances in catalytic transformation of biomass-derived 5-hydroxymethylfurfural into the innovative fuels and chemicals. *Renew. Sustain. Energy Rev.* **2017**, *74*, 230–257. [CrossRef]
12. Kucherov, F.A.; Romashov, L.V.; Galkin, K.I.; Ananikov, V.P. Chemical Transformations of Biomass-Derived C6-Furanic Platform Chemicals for Sustainable Energy Research, Materials Science, and Synthetic Building Blocks. *ACS Sustain. Chem. Eng.* **2018**, *6*, 8064–8092. [CrossRef]
13. Xia, H.; Xu, S.; Hu, H.; An, J.; Li, C. Efficient conversion of 5-hydroxymethylfurfural to high-value chemicals by chemo- and bio-catalysis. *RSC Adv.* **2018**, *8*, 30875–30886. [CrossRef]
14. Werpy, T.; Petersen, G. *Top Value Added Chemicals from Biomass: Volume I—Results of Screening for Potential Candidates from Sugars and Synthesis Gas*; National Renewable Energy Lab.: Golden, CO, USA, 2004.
15. Papageorgiou, G.Z.; Papageorgiou, D.G.; Terzopoulou, Z.; Bikiaris, D.N. Production of bio-based 2,5-furan dicarboxylate polyesters: Recent progress and critical aspects in their synthesis and thermal properties. *Eur. Polym. J.* **2016**, *83*, 202–229. [CrossRef]
16. Gandini, A.; Silvestre, A.J.D.; Neto, C.P.; Sousa, A.F.; Gomes, M. The furan counterpart of poly (ethylene terephthalate): An alternative material based on renewable resources. *J. Polym. Sci. Part A Polym. Chem.* **2009**, *47*, 295–298. [CrossRef]
17. Eerhart, A.J.J.E.; Faaij, A.P.C.; Patel, M.K. Replacing fossil based PET with biobased PEF; process analysis, energy and GHG balance. *Energy Environ. Sci.* **2012**, *5*, 6407–6422. [CrossRef]
18. Partenheimer, W.; Grushin, V.V. Synthesis of 2,5-Diformylfuran and Furan-2,5-Dicarboxylic Acid by Catalytic Air-Oxidation of 5-Hydroxymethylfurfural. Unexpectedly Selective Aerobic Oxidation of Benzyl Alcohol to Benzaldehyde with Metal=Bromide Catalysts. *Adv. Synth. Catal.* **2001**, *343*, 102–111. [CrossRef]
19. Singh, S.K. Heterogeneous Bimetallic Catalysts for Upgrading Biomass-Derived Furans. *Asian J. Org. Chem.* **2018**, *7*, 1901–1923. [CrossRef]
20. Cattaneo, S.; Stucchi, M.; Villa, A.; Prati, L. Gold Catalysts for the Selective Oxidation of Biomass-Derived Products. *ChemCatChem* **2019**, *11*, 309–323. [CrossRef]
21. Ventura, M.; Dibenedetto, A.; Aresta, M. Heterogeneous catalysts for the selective aerobic oxidation of 5-hydroxymethylfurfural to added value products in water. *Inorg. Chim. Acta* **2018**, *470*, 11–21. [CrossRef]

22. Pasini, T.; Piccinini, M.; Blosi, M.; Bonelli, R.; Albonetti, S.; Dimitratos, N.; Lopez-Sanchez, J.A.; Sankar, M.; He, Q.; Kiely, C.J.; et al. Selective oxidation of 5-hydroxymethyl-2-furfural using supported gold–copper nanoparticles. *Green Chem.* **2011**, *13*, 2091–2099. [CrossRef]
23. Albonetti, S.; Pasini, T.; Lolli, A.; Blosi, M.; Piccinini, M.; Dimitratos, N.; Lopez-Sanchez, J.A.; Morgan, D.J.; Carley, A.F.; Hutchings, G.J.; et al. Selective oxidation of 5-hydroxymethyl-2-furfural over TiO_2-supported gold–copper catalysts prepared from preformed nanoparticles: Effect of Au/Cu ratio. *Catal. Today* **2012**, *195*, 120–126. [CrossRef]
24. Lolli, A.; Albonetti, S.; Utili, L.; Amadori, R.; Ospitali, F.; Lucarelli, C.; Cavani, F. Insights into the reaction mechanism for 5-hydroxymethylfurfural oxidation to FDCA on bimetallic Pd–Au nanoparticles. *Appl. Catal. A Gen.* **2015**, *504*, 408–419. [CrossRef]
25. Megías-Sayago, C.; Lolli, A.; Bonincontro, D.; Penkova, A.; Albonetti, S.; Cavani, F.; Odriozola, J.A.; Ivanova, S. Effect of gold particles size on Au/C catalyst selectivity in HMF oxidation reaction. *ChemCatChem* **2019**. [CrossRef]
26. Villa, A.; Schiavoni, M.; Campisi, S.; Veith, G.M.; Prati, L. Pd-modified Au on carbon as an effective and durable catalyst for the direct oxidation of HMF to 2,5-furandicarboxylic acid. *ChemSusChem* **2013**, *6*, 609–612. [CrossRef]
27. Wan, X.; Zhou, C.; Chen, J.; Deng, W.; Zhang, Q.; Yang, Y.; Wang, Y. Base-Free Aerobic Oxidation of 5-Hydroxymethyl-furfural to 2,5-Furandicarboxylic Acid in Water Catalyzed by Functionalized Carbon Nanotube-Supported Au–Pd Alloy Nanoparticles. *ACS Catal.* **2014**, *4*, 2175–2185. [CrossRef]
28. Bonincontro, D.; Lolli, A.; Villa, A.; Prati, L.; Dimitratos, N.; Veith, G.M.; Chinchilla, L.E.; Botton, G.A.; Cavani, F.; Albonetti, S. AuPd-nNiO as an effective catalyst for the base-free oxidation of HMF under mild reaction conditions. *Green Chem.* **2019**, *21*, 4090–4099. [CrossRef]
29. Gupta, N.K.; Nishimura, S.; Takagaki, A.; Ebitani, K. Hydrotalcite-supported gold-nanoparticle-catalyzed highly efficient base-free aqueous oxidation of 5-hydroxymethylfurfural into 2,5-furandicarboxylic acid under atmospheric oxygen pressure. *Green Chem.* **2011**, *13*, 824–827. [CrossRef]
30. Ferraz, C.P.; Zieliński, M.; Pietrowski, M.; Heyte, S.; Dumeignil, F.; Rossi, L.M.; Wojcieszak, R. Influence of Support Basic Sites in Green Oxidation of Biobased Substrates Using Au-Promoted Catalysts. *ACS Sustain. Chem. Eng.* **2018**, *6*, 16332–16340. [CrossRef]
31. Dessbesell, L.; Souzanchi, S.; Rao, K.T.V.; Carrillo, A.A.; Bekker, D.; Hall, K.A.; Lawrence, K.M.; Tait, C.L.J.; Xu, C. (Charles) Production of 2,5-furandicarboxylic acid (FDCA) from starch, glucose, or high-fructose corn syrup: Techno-economic analysis. *Biofuels Bioprod. Biorefin.* **2019**, *13*, 1234–1245. [CrossRef]
32. Wiles, C.; Watts, P. Continuous process technology: A tool for sustainable production. *Green Chem.* **2013**, *16*, 55–62. [CrossRef]
33. Morse, P.D.; Beingessner, R.L.; Jamison, T.F. Enhanced Reaction Efficiency in Continuous Flow. *Isr. J. Chem.* **2017**, *57*, 218–227. [CrossRef]
34. Paired Electrochemical Oxidation process for feasible industrial production of the crucial FDCA building block for the bioplastic industry | PairElOx Project | H2020 | CORDIS | European Commission. Available online: https://cordis.europa.eu/project/rcn/217556/factsheet/en (accessed on 19 November 2019).
35. Kastl, R.; Kaufmann, E.; Bodenmüller, A.; Sedelmeier, G. Process for Preparing 2,5-Furandicarboxylic Acid. assigned to SYNPHABASE AG [CH/CH]; Gueterstrasse 82 4133 Pratteln (CH). Patent WO2019/072920, 18 April 2019.
36. Liguori, F.; Barbaro, P.; Calisi, N. Continuous-Flow Oxidation of HMF to FDCA by Resin-Supported Platinum Catalysts in Neat Water. *ChemSusChem* **2019**, *12*, 2558–2563. [CrossRef] [PubMed]
37. Latsuzbaia, R.; Bisselink, R.; Anastasopol, A.; Van der Meer, H.; Van Heck, R.; Yagüe, M.S.; Zijlstra, M.; Roelands, M.; Crockatt, M.; Goetheer, E.; et al. Continuous electrochemical oxidation of biomass derived 5-(hydroxymethyl)furfural into 2,5-furandicarboxylic acid. *J. Appl. Electrochem.* **2018**, *48*, 611–626. [CrossRef]
38. Drioli, E.; Brunetti, A.; Profio, G.D.; Barbieri, G. Process intensification strategies and membrane engineering. *Green Chem.* **2012**, *14*, 1561–1572. [CrossRef]
39. Brunetti, A.; Zito, P.F.; Giorno, L.; Drioli, E.; Barbieri, G. Membrane reactors for low temperature applications: An overview. *Chem. Eng. Process. Process. Intensif.* **2018**, *124*, 282–307. [CrossRef]
40. Xue, J.; Wu, T.; Dai, Y.; Xia, Y. Electrospinning and Electrospun Nanofibers: Methods, Materials, and Applications. *Chem. Rev.* **2019**, *119*, 5298–5415. [CrossRef]

41. Gualandi, C.; Celli, A.; Zucchelli, A.; Focarete, M.L. Nanohybrid Materials by Electrospinning. In *Organic-Inorganic Hybrid Nanomaterials*; Advances in Polymer Science; Kalia, S., Haldorai, Y., Eds.; Springer International Publishing: Cham, Switzerland, 2015; pp. 87–142. ISBN 978-3-319-13593-9.
42. Li, B.; Zhang, B.; Nie, S.; Shao, L.; Hu, L. Optimization of plasmon-induced photocatalysis in electrospun Au/CeO2 hybrid nanofibers for selective oxidation of benzyl alcohol. *J. Catal.* **2017**, *348*, 256–264. [CrossRef]
43. Liu, Y.; Chen, H.-S.; Li, J.; Yang, P. Morphology adjustment of one dimensional CeO2 nanostructures via calcination and their composite with Au nanoparticles towards enhanced catalysis. *RSC Adv.* **2015**, *5*, 37585–37591. [CrossRef]
44. Hao, Y.; Shao, X.; Li, B.; Hu, L.; Wang, T. Mesoporous TiO2 nanofibers with controllable Au loadings for catalytic reduction of 4-nitrophenol. *Mater. Sci. Semicond. Process.* **2015**, *40*, 621–630. [CrossRef]
45. Yue, G.; Li, S.; Li, D.; Liu, J.; Wang, Y.; Zhao, Y.; Wang, N.; Cui, Z.; Zhao, Y. Coral-like Au/TiO$_2$ Hollow Nanofibers with Through-Holes as a High-Efficient Catalyst through Mass Transfer Enhancement. *Langmuir* **2019**, *35*, 4843–4848. [CrossRef]
46. Liu, Y.; Jiang, G.; Li, L.; Chen, H.; Huang, Q.; Jiang, T.; Du, X.; Chen, W. Preparation of Au/PAN nanofibrous membranes for catalytic reduction of 4-nitrophenol. *J. Mater. Sci.* **2015**, *50*, 8120–8127. [CrossRef]
47. Liu, Y.; Zhang, K.; Li, W.; Ma, J.; Vancso, G.J. Metal nanoparticle loading of gel-brush grafted polymer fibers in membranes for catalysis. *J. Mater. Chem. A* **2018**, *6*, 7741–7748. [CrossRef]
48. Cipriani, E.; Zanetti, M.; Bracco, P.; Brunella, V.; Luda, M.P.; Costa, L. Crosslinking and carbonization processes in PAN films and nanofibers. *Polym. Degrad. Stab.* **2016**, *123*, 178–188. [CrossRef]
49. Mathur, R.B.; Bahl, O.P.; Sivaram, P. Thermal degradation of polyacrylonitrile fibres. *Curr. Sci.* **1992**, *62*, 662–669.
50. Albonetti, S.; Lolli, A.; Morandi, V.; Migliori, A.; Lucarelli, C.; Cavani, F. Conversion of 5-hydroxymethylfurfural to 2,5-furandicarboxylic acid over Au-based catalysts: Optimization of active phase and metal–support interaction. *Appl. Catal. B Environ.* **2015**, *163*, 520–530. [CrossRef]
51. Maccaferri, E.; Mazzocchetti, L.; Benelli, T.; Zucchelli, A.; Giorgini, L. Morphology, thermal, mechanical properties and ageing of nylon 6,6/graphene nanofibers as Nano$_2$ materials. *Compos. Part B Eng.* **2019**, *166*, 120–129. [CrossRef]

© 2020 by the authors. Licensee MDPI, Basel, Switzerland. This article is an open access article distributed under the terms and conditions of the Creative Commons Attribution (CC BY) license (http://creativecommons.org/licenses/by/4.0/).

Article

Microemulsion vs. Precipitation: Which Is the Best Synthesis of Nickel–Ceria Catalysts for Ethanol Steam Reforming?

Cristina Pizzolitto [1], Federica Menegazzo [1], Elena Ghedini [1], Arturo Martínez Arias [2], Vicente Cortés Corberán [2] and Michela Signoretto [1,*]

[1] CATMAT Lab, Department of Molecular Sciences and Nanosystems, Ca' Foscari University Venice and INSTM RU of Venice, via Torino 155, I-30172 Venezia Mestre, Italy; cristina.pizzolitto@unive.it (C.P.); fmenegaz@unive.it (F.M.); gelena@unive.it (E.G.)
[2] Institute of Catalysis and Petroleum Chemistry (ICP), CSIC, Calle Marie Curie 2, 28049 Madrid, Spain; amartinez@icp.csic.es (A.M.A.); vcortes@icp.csic.es (V.C.C.)
* Correspondence: miky@unive.it

Abstract: Ethanol steam reforming is one of the most promising ways to produce hydrogen from biomass, and the goal of this research is to investigate robust, selective and active catalysts for this reaction. In particular, this work is focused on the effect of the different ceria support preparation methods on the Ni active phase stabilization. Two synthetic approaches were evaluated: precipitation (with urea) and microemulsion. The effects of lanthanum doping were investigated too. All catalysts were characterized using N_2-physisorption, temperature programmed reduction (TPR), XRD and SEM, to understand the influence of the synthetic approach on the morphological and structural features and their relationship with catalytic properties. Two synthesis methods gave strongly different features. Catalysts prepared by precipitation showed higher reducibility (which involves higher oxygen mobility) and a more homogeneous Ni particle size distribution. Catalytic tests (at 500 °C for 5 h using severe Gas Hourly Space Velocity conditions) revealed also different behaviors. Though the initial conversion (near complete) and H_2 yield (60%, i.e., 3.6 mol H_2/mol ethanol) were the same, the catalyst prepared by microemulsion was deactivated much faster. Similar trends were found for La-promoted supports. Catalyst deactivation was mainly related to coke deposition as was shown by SEM of the used samples. Higher reducibility of the catalysts prepared by the precipitation method led to a decrease in coke deposition rate by facilitating the removal of coke precursors, which made them the more stable catalysts of the reaction.

Keywords: ethanol steam reforming; Ni/CeO_2; microemulsion; coke resistance; lanthanum doping

1. Introduction

Hydrogen, one of the most useful intermediate products, can be the ideal candidate to solve the environmental problems [1]. Although, it is not a fuel by itself, it is considered as the future energy vector owing to its great potential for generating electricity by fuel cells [2]. A critical issue in the use of hydrogen for energy applications is its production method. In fact, despite hydrogen being the most abundant element in the universe, it does not exist in significant amounts in its elemental form [3,4]. Therefore, it must be produced from other sources. Nowadays, 96% of the hydrogen produced worldwide derives from the conversion of fossil resources [5], mainly from natural gas by steam reforming. Nevertheless, one potential for the future is the possibility of hydrogen generation from renewable sources. Among the most attractive processes, the steam reforming of light alcohols such as methanol and ethanol plays a key role [6]. Indeed, methanol can be produced by syngas derived from biomass, while ethanol can be generated by fermentation of carbohydrate sources [7]. In addition to the said use of renewable raw materials, the use of alcohol as the main resource for hydrogen production has many advantages such as low cost, easy transportation in liquid form and the possibility of its conversion to hydrogen in

Citation: Pizzolitto, C.; Menegazzo, F.; Ghedini, E.; Martínez Arias, A.; Cortés Corberán, V.; Signoretto, M. Microemulsion vs. Precipitation: Which Is the Best Synthesis of Nickel-Ceria Catalysts for Ethanol Steam Reforming? *Processes* **2021**, *9*, 77. https://doi.org/10.3390/pr9010077

Received: 13 November 2020
Accepted: 26 December 2020
Published: 31 December 2020

Publisher's Note: MDPI stays neutral with regard to jurisdictional claims in published maps and institutional affiliations.

Copyright: © 2020 by the authors. Licensee MDPI, Basel, Switzerland. This article is an open access article distributed under the terms and conditions of the Creative Commons Attribution (CC BY) license (https://creativecommons.org/licenses/by/4.0/).

relatively mild reaction conditions [8] with highly efficient and cost-effective processes. For example, conversion of ethanol into hydrogen via steam reforming provides six moles of hydrogen per mole of ethanol because it can extract hydrogen not only from ethanol but also from water ($CH_3CH_2OH + 3H_2O = 2CO_2 + 6H_2$) [9–11]. However, ethanol steam reforming (ESR) follows a complex reaction pathway, summarized in Figure 1. As may be seen, several by-products such as carbon monoxide, methane, ethylene, acetaldehyde and more complex carbon species can be generated under reaction conditions [12]. For this reason, the catalyst formulation is not a trivial task: it should be properly formulated to be functional to direct the reaction to maximize hydrogen yield and, at the same time, to suppress the unwanted side reactions. Common catalysts for ESR are metals, such as Pt, Pd, Rh, Ni, Co and Cu, [13,14] supported on different oxides, mainly Al_2O_3, SiO_2, CeO_2, ZrO_2, TiO_2, MgO and La_2O_3 [15–17]. Among them, nickel is an attractive active phase for its low cost and high activity, comparable to that of noble metals. In addition, ceria is an interesting support since it belongs to the partially reducible oxides (PROs) [8]. Indeed, it is commonly used in different oxidation reactions such as CO oxidation [18], preferential oxidation (PROX) of CO for hydrogen purification [19], water gas shift (WGS) reaction, as well as oxygen-conducting membranes for solid oxide fuel cells and many other processes [20,21]. Thanks to the redox ability and strong interaction with nickel, ceria has been extensively studied in the reforming field [22]. Ceria redox ability can be modulated by a careful control of structural defects [23–25]: the higher the number of defective sites, the more effective the redox pump. Therefore, lanthanum oxide has been added as promoter due to its possible substitution as La^{3+} in the Ce^{4+} lattice [18] which may lead to the formation of defective sites.

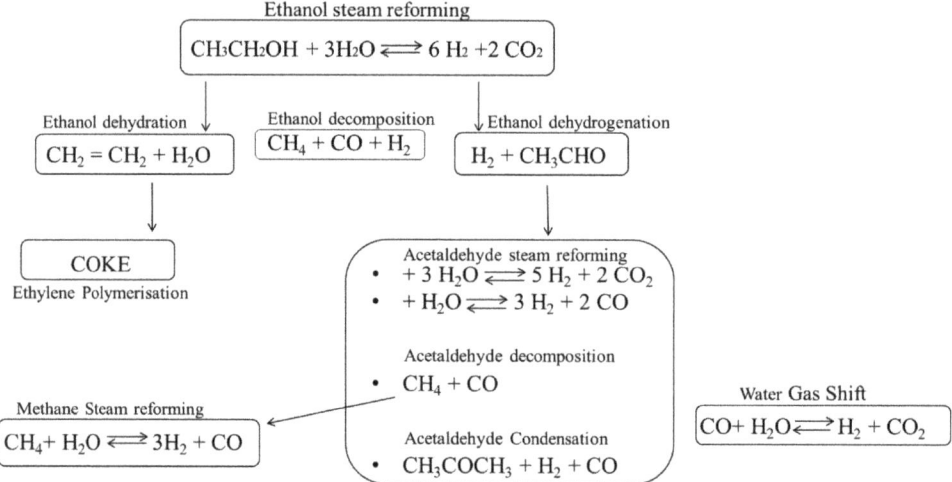

Figure 1. Reaction pathways involved in the ethanol steam reforming (ESR) process.

This work has been focused on the preparation method of the support for nickel–ceria-based catalysts. Two different synthetic methods have been investigated for ceria supports preparation: precipitation and reverse microemulsion. Precipitation is the standard approach used for metal oxides preparation. With this method, however, it is difficult to control particle size distribution. On the contrary, reverse microemulsion can be an innovative way to modulate and control the textural properties of new materials. This approach is based on the formation of nanospherical micelles inside which the precipitation of the oxide takes place. In this way, as reported by Eriksson et al. [26], a suitable environment for producing small nanoparticles with narrow size distribution can be generated. Accordingly, the motivation of this work is to focus on the investigation of the influence of the preparation method on the activity, stability and regenerability of nickel–ceria-based catalysts for hydrogen production via ESR, that was investigated under severe Gas Hourly Space Velocity (GHSV) conditions. To the best of our knowledge this is the first use in ESR of Ni supported on La doped CeO_2 prepared by reverse microemulsion. To achieve this goal, different characterization approaches were used. In particular, the correlation between the synthetic method and structural and morphological properties was investigated by N_2-physisorption, SEM, XRD, and H_2-temperature programmed reduction (TPR). The effects of lanthanum doping have been investigated too.

2. Experimental Part

2.1. Support Preparation

Precipitation method: the supports (hereinafter denoted as Ce P and CeLa P) were synthesized by precipitation with urea at 100 °C with aqueous solutions of $(NH_4)_2Ce(NO_3)_6$ (Sigma Aldrich, St Louis, MO, USA) and additionally $La(NO_3)_3 \cdot 6H_2O$, in the adequate amount to obtain a 5 wt % of lanthanum(Sigma Aldrich, St Louis, MO, USA); in the final sample, for the latter. The solution was mixed and boiled at 100 °C for 6 h. The precipitates were washed with deionized water and dried at 110 °C for 18 h. The material was then calcined under air flow (30 mL/min) at 650 °C for 3 h.

Reverse microemulsion method: the supports (hereinafter denoted as Ce M and CeLa M) were synthesized by preparing two different microemulsion systems: (A) Saline microemulsion composed by: 450 mL n-heptane 99% (Sigma Aldrich, GMBH, Riedstr, Germany), 90 mL triton X(Sigma Aldrich, St Louis, USA, MO, USA), 92 mL 1-hexanol 98% (Sigma Aldrich, St Louis, MO, USA), 50 mL aqueous solution of $Ce(NO_3)_3 \cdot 6H_2O$ (0.5 M); (B) Basic microemulsion composed by 450 mL n-heptane 99%, 1-hexanol 98%, 90 mL Triton X-100 and 50 mL basic solution (0.5 M NaOH). The proper amount of lanthanum precursor was added to obtain a 5 wt % of lanthanum in the final sample. Both microemulsions were stirred at 100 rpm for 1 h. Then, solution B was added to solution A and was kept under stirring at 100 rpm for 24 h. The precipitates were separated by centrifugation, washed with methanol, dried at 120 °C for 18 h and calcined in air at 650 °C for 3 h.

2.2. Catalyst Preparation

Nickel was introduced on the supports by incipient wetness impregnation with a proper amount of $Ni(NO_3)_2 \cdot 6H_2O$ aqueous solution to obtain 8 wt % of nickel on the catalysts. After drying at 110 °C for 18 h, calcination was performed in air at 650 °C for 4 h.

2.3. Catalysts Characterization

Specific surface areas and pore size distributions were evaluated by N_2 adsorption/desorption isotherms at −196 °C using a Tristal II Plus Micromeritics. (Micromeritics, Milan, Italy); The surface area was calculated using the Brunauer–Emmett–Teller (BET) Equation [27] method while pore size distribution was determined by the BJH (Barrett, Joyner, and Halenda) method [28], applied to the N_2 desorption branch of the isotherm.

The Ni and La contents were determined by atomic absorption spectroscopy (AAS) after microwave disaggregation of the samples (100 mg), using a Perkin-Elmer Analyst (Perkin-Elmer, Waltham, MA, USA); 100 spectrometer.

The morphology of the catalysts was studied by scanning electron microscopy (SEM) with a table-top Hitachi instrument, model TM-1000 (Hitachi, Ramsey, NJ, USA), after depositing the ground powder sample on a double-sided lacey carbon ribbon.

X-ray diffractograms were obtained on a Seifert XRD 3000P diffractometer; using nickel-filtered Cu Kα radiation operating at 40 kV and 40 mA, using a 0.02° step size and 2 s counting time per step. Analysis of the diffraction peaks was conducted with the software ANALYZE Rayflex Version 2.293.

Temperature programmed reduction (TPR) measurements were carried out using lab-made equipment: samples (100 mg) were heated with a temperature rate of 10 °C/min from 25 °C to 900 °C in a 5% H_2/He flow (40 mL/min). The effluent gases were analyzed by a Thermal Conductivity Detector (GOW-MAC InstrumentCo., Shannon, Irland)

2.4. Catalytic Tests

Catalysts were tested for ESR at 500 °C and atmospheric pressure, charging the feed with molar composition of water:ethanol:He = 18.4:3.1:78.5 and W/F = 0.12 g_{cat}.h/mol ethanol, in a stainless steel, fixed bed tubular reactor placed in an equipment Microactivity Reference model MAXXXM3-(PID Eng and Tech, Madrid, Spain). Prior to the reaction, fresh catalyst samples were activated under flow of 10% O_2 in He at 650 °C for 1 h. Catalytic stability tests were conducted at 500 °C for 5 h. After the first run, the catalyst was cooled down and flushed under inert flow, and then reactivated using the same procedure of the initial activation, heating up to 650 °C at 10 °C/min and keeping this temperature for 1 h, under a flow of 10% O_2 in He. After cooling down to 500 °C in inert flow, a second run was conducted with the regenerated samples under the same condition of the first run. Tests for each sample were reproducible within experimental error. Reactants and products were analyzed online by GC on a Varian Star 3400 CX instrument (Varian, Cridersville, OH, USA); equipped with two columns, molecular sieve and Porapak Q and the detector of thermal conductivity. After the analysis, conversion of ethanol and hydrogen yield were calculated as follows:

Conversion of ethanol:

$$conversion\ (\%) = \left[\frac{n_{in}(EtOH) - n_{out}(EtOH)}{n_{in}(EtOH)}\right] \times 100 \quad (1)$$

H_2 yield:

$$yield\ (\%) = \frac{fH2\ out}{6 \times fEtOHin} \times 100 \quad (2)$$

with n number of moles; f, flux in mL/min.

3. Results and Discussion

3.1. Catalysts Characterization

The specific surface area is one of the most important parameters in the design of a heterogeneous catalyst: a high surface area greatly improves the dispersion of the active phase [29,30]. Figure 2 shows N_2-physisorption isotherms of the samples, while the calculated values of specific surface area, mean pore radius and pore volume are reported in Table 1.

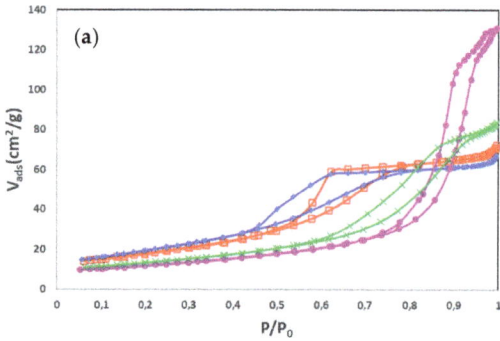
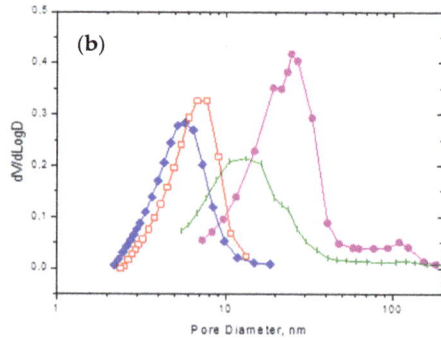

Figure 2. N_2-physisorption isotherms of catalysts (**a**) and their pore size distributions (**b**): NiCe P (red squares); NiCe M (green cross); NiLaCe P (blue rhombs); NiLaCe M (violet circle).

Table 1. Physico-chemical properties of the catalysts.

Samples	S_{BET} [a] (cm^3/g)	Mean Pore Diameter [b] (nm)	V_{pore} [c] (cm^3/g)	NiO Mean Particle Size [d] (nm)
NiCe P	64	7.0	0.11	27
NiLaCe P	71	6.8	0.12	25
NiCe M	48	10.5	0.12	15
NiLaCe M	42	12.0	0.16	14

[a] Specific surface area calculated via BET; [b] average pore diameter, and [c] pore volume calculated via BJH method; [d] calculated by Scherrer Equation.

All samples exhibited the IV-type isotherm that is typical of mesoporous materials according to IUPAC classification [31]. However, the shapes and hysteresis loops of the isotherms were quite different. The hysteresis loop of precipitated samples is H2 type, characteristic of solids with pores of irregular shape and dimension. Conversely, the catalysts obtained by the microemulsion approach show a hysteresis profile more difficult to classify being a combination of H2 and H3 hysteresis loops associated with a complex pores structure. H3 loops are, generally, associated with non-rigid aggregates of plate-like particles (e.g., certain clays) or with a pore network consisting of macropores that are not completely filled with pore condensate [32]. The pore size distribution obtained by BJH is consistent with the N_2 adsorption-desorption profiles, in fact the pore size distribution for the "M" type catalysts was broader and at higher values, at the limit of macroporosity, than for LaCe P and NiLaCe P samples. Moreover, both precipitated samples exhibit a higher BET surface area than the catalysts obtained by the microemulsion approach (Table 1).

Analytical Ni amount was the same for all the samples (7.5 wt %), only slightly lower than the nominal value of 8 wt %. As for La amount, it was around 5 wt %. The particles morphology and catalyst size were determined using microscopy techniques. Figure 3 shows SEM images of the fresh samples. As can be seen, the appearance of the materials prepared by different techniques was notably different. Samples synthesized by precipitation were made of agglomerated spherical particles of 1.8–2 μm, while catalysts prepared by microemulsion presented a wrinkled surface covered by small superficial cubic particles.

XRD analyses using Scherrer refinement were carried out to determine the crystal size of the support and the metal phases in the samples. Figure 4 compares the XRD patterns of the four NiCe samples prepared via different support preparation methods.

As for the fresh samples, XRD profiles showed a fluorite-type phase of ceria with characteristic reflections at 2θ = 28°, 33°, 47°, 56°, 59°, and 69° associated with (111), (200), (220), (311), (222) and (400) planes of the cubic phase, respectively [33]. No diffraction lines related to lanthanum nor lanthanum oxide can be evidenced in XRD spectra, despite its almost 5 wt % loading. This could be reasonably due to the incorporation of La^{3+} ions in the ceria lattice [18]. A mean size of 11 and 9 nm was calculated by Sherrer for the crystal particle size of CeO_2, respectively, in NiCeP and NiCeM. La addition does not significantly affect ceria size. Regarding the active phase, the occurrence of NiO was clearly detected, with the characteristic diffraction lines at 2θ 37° and 43.4° [34]. The presence of nickel in its oxidic form was not unexpected, because the analyses have been performed on calcined catalysts, as the samples charged in the reactor for catalytic ESR testing. Table 1 shows the crystal size of NiO, calculated from the analysis of the most intense diffractions, corresponding to 2θ = 43.4°. The patterns of the precipitated samples showed sharper and more intense diffraction lines, meaning that the particles of ceria and NiO were bigger and more crystalline than for the catalysts prepared via microemulsion (Table 1).

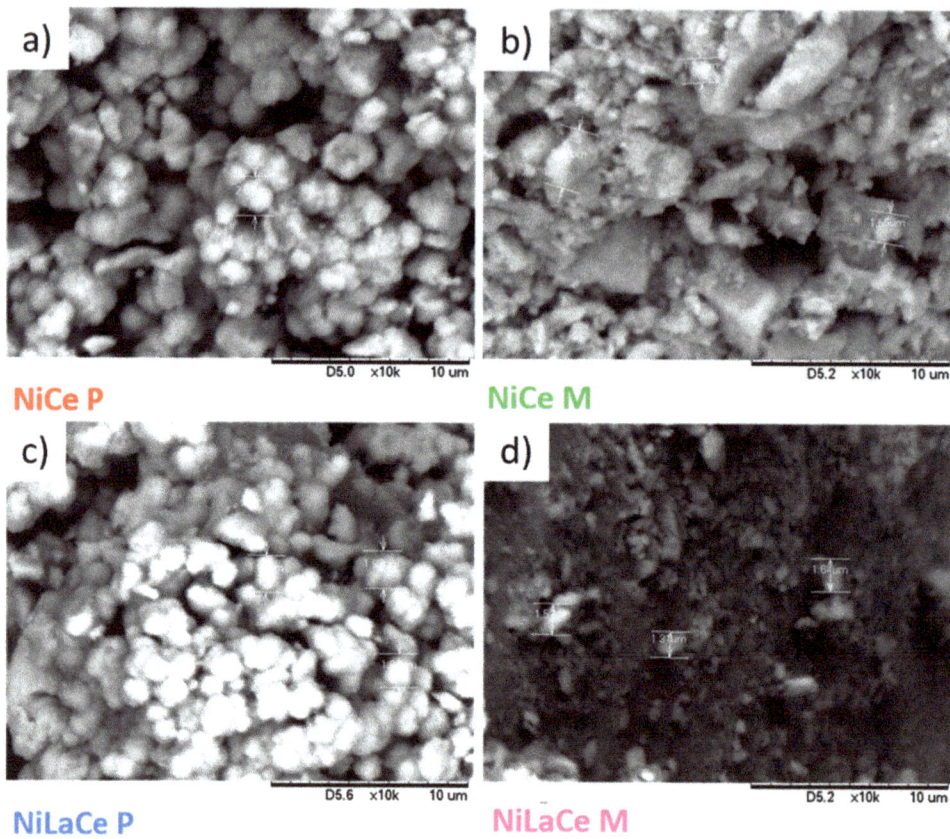

Figure 3. Representative SEM images of fresh catalysts (**a**) NiCe P, (**b**) NiCe M, (**c**) NiLaCe P and (**d**) NiLaCe M.

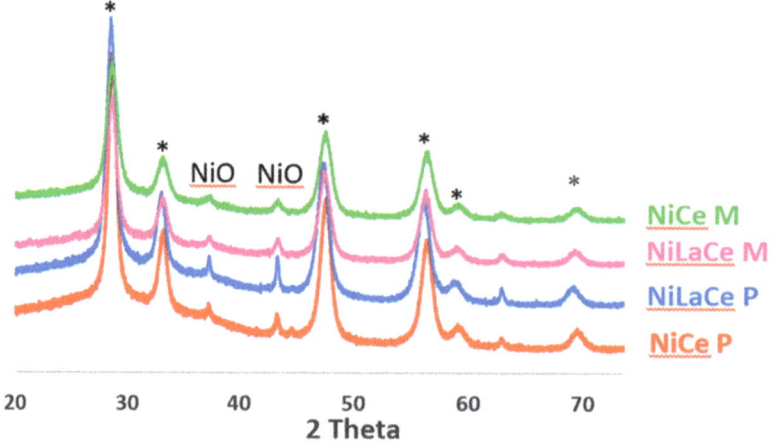

Figure 4. XRD patterns of fresh catalysts after calcination. (* denoted peaks of ceria crystal phases.)

These preliminary analyses indicated that the support preparation method strongly affected morphological and structural features of the final catalysts. Therefore, further characterizations were performed. TPR technique was used to identify the different NiO species on the ceria surface and their reduction features and to determine the support reduction temperature. The TPR profiles are reported in Figure 5. The most evident difference between the TPR profiles of the two different techniques is the number of NiO species interacting with the support. The profiles of samples prepared by precipitation presented three broad reduction peaks at 204 °C, 249 °C and 329 °C that can be associated with NiO reduction. On the contrary, the TPR curves of samples synthesized by microemulsion showed only one sharp peak centered at 470 °C. Therefore, TPR analyses clearly showed that the synthetic approach has a deep effect on Ni reducibility, which could affect their catalytic behavior. NiO was reduced at temperatures below 400 °C for samples prepared by precipitation, while for catalysts synthesized by microemulsion the temperature needed for NiO reduction is at least 400 °C. This difference can be reasonably ascribable to the different metal–support interactions that can be formed during the oxide precipitation: the stronger the interaction, the higher the reduction temperature. The broad maxima at higher temperatures are related to the support, since it is known that the ceria can be reduced from Ce^{4+} to Ce^{3+} at temperatures above 700 °C [35]. As reported in the literature, one can envisage, for the precipitated supports, one small and broad peak at 800 °C, while for the support synthesized by microemulsion, there are two small and broad peaks at 750 and 1000 °C, respectively.

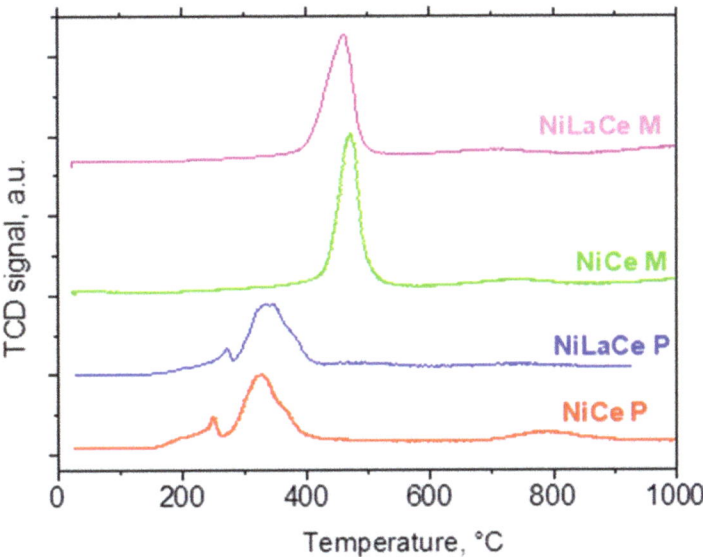

Figure 5. Temperature programmed reduction (TPR) profiles of catalysts.

From all these characterization results, remarkable differences between two methodologies have arisen. Precipitation results in slightly higher surface areas, well-defined spherical-shaped particles and high crystallinity. At the same time, catalysts prepared by this technique presented a weaker metal support interaction, and a consequent easier metal reducibility. In contrast, NiCe M and NiLaCe M showed smaller NiO dimensions and only one stronger NiO interaction with the support, which could strongly affect NiO reducibility in reaction conditions. The effect of lanthanum addition is minimal with respect the difference in the preparation methodology. Consequently, it can be affirmed that the support preparation method strongly influences morphological, structural, and chemical properties of the final catalysts.

3.2. Catalytic Performances

ESR catalytic tests were carried out at 500 °C and atmospheric pressure using severe GHSV conditions. Figure 6 reports ethanol conversion (straight line) and hydrogen yield (dotted lines) along 5 h of time on stream. Both NiCe P and NiCe M catalysts presented a high initial activity with nearly complete ethanol conversion and 60% of hydrogen yield, considering that NiCe M has a slightly lower performance than NiCe P. This could be due to a more difficult reducibility of Ni in the sample prepared by microemulsion, as was demonstrated by TPR analyses. In fact, it should be noted that these materials were not reduced before the reaction test and it is relevant to verify the reducing power of ethanol, confirming what was previously demonstrated by Pizzolitto et al. [18].

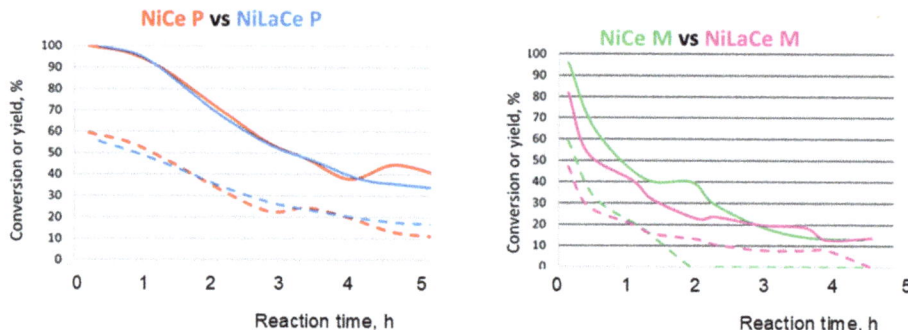

Figure 6. Catalytic activity in the ethanol steam reforming on NiLaCe P (**left**) and NiLaCe M (**right**) catalysts in comparison with the non-doped samples: ethanol conversion (full line) and hydrogen yield (dotted line).

Nevertheless, both the catalysts suffered a progressive deactivation over time on stream. However, the deactivation presented a different degree: after 2.5 h, NiCe M completely lost its activity for hydrogen production, maintaining at the same time a very low ethanol conversion. The catalyst prepared via precipitation kept 50% conversion and 30% hydrogen yield after 5 h on stream. The behavior of lanthanum-doped samples was very similar to the corresponding non-doped catalysts under the tested reaction conditions.

Characterizations of catalyst samples recovered after reaction were also performed to understand more deeply the evolution of catalytic behavior. Examples of SEM images of used samples are shown in Figure 7. As can be seen, the catalyst prepared via microemulsion, NiCe M, was almost completely covered by carbon, while the sample prepared by precipitation, NiCe P, still showed a very clear surface despite the fact that it had some dark agglomerates associated with carbon deposits. Although Carbon was not formed as nanotubes or nanowires, it was in a more compact form, either polymeric or graphitic. These results perfectly matched with the catalytic results: the complete activity loss for the NiCe M sample is due to the complete coverage of active sites by carbon. On the contrary, only a small portion of metals was covered by coke deposits, and therefore the sample was still active after 5 h of reaction.

Figure 7. SEM images of used catalyst samples recovered after ethanol steam reforming tests: (**a**) Used NiCe P and (**b**) Used NiCe M.

To further understand the reasons of this discrepancy in catalytic behavior, X-ray diffractograms of the used catalysts were obtained. Figure 8 compares the XRD patterns of fresh and used catalysts from both preparation methods. The patterns of used samples presented reflections attributed to carbon species at 2θ 26.4° [36] probably with a graphite-like structure. For the used catalysts, the crystal particle size of CeO_2 slightly increased (11 and 10 nm, respectively, for used NiCe P and NiCe M) and the reflections attributed to metallic Ni appear at 2θ of 44.5° and 51.8°, while those of NiO at 43.4° disappear (see inset in Figure 8). This evidenced that the reaction mixture, that is the reactant ethanol, allowed the reduction in the metal phase, thus activating the catalysts for the reaction.

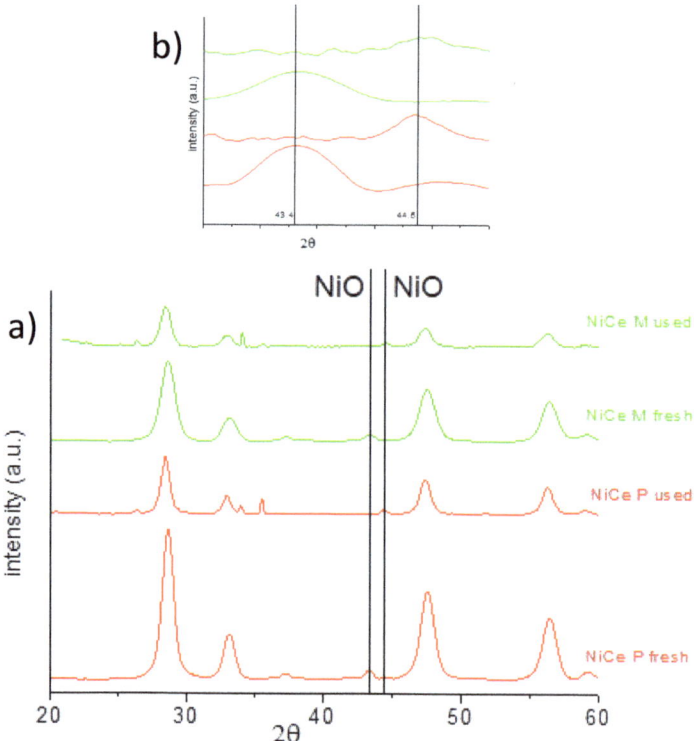

Figure 8. (a) XRD patterns of fresh and used catalysts after calcination. (Reflections at 2θ 35° are due to remains of SiC used as diluent in the catalytic bed.) and (b) magnification on reflections of NiO at 2θ 43.3 and 44.5°.

Moreover, to determine the possibility of reusability of the catalysts, reactivation of used catalysts was carried out, followed by a second run of the catalytic test. Figure 9 compares ethanol conversion and hydrogen yield of fresh and used catalysts for both undoped materials. After the regeneration in air, the initial H_2 yield on NiCe P, the most active one, was lower than that of the fresh sample, decreasing to 17% after 5 h of reaction (instead of 30% of the first run). Therefore, catalyst regeneration did not allow its complete reactivation, indicating either possible sintering of the catalyst or incomplete removal of the carbonaceous deposits. In the case of NiCe M, it was completely reactivated after the regeneration step, as the curve of ethanol conversion of the regenerated sample practically overlapped with that of the fresh one. However, considering hydrogen yield, an even faster deactivation is visible, with no hydrogen production after just 2 h. This apparent discrepancy, of the same conversion but different yield, is due to a different selectivity, indicating that some changes in the nature of the material occurred.

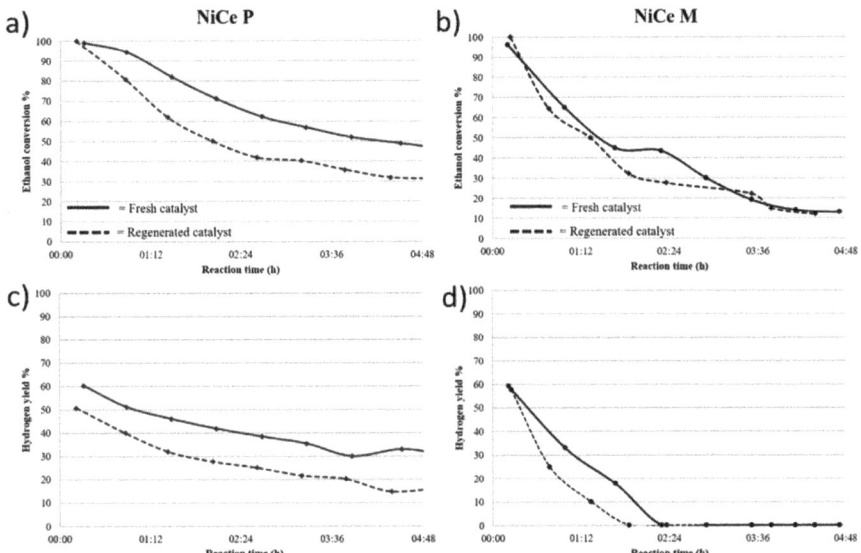

Figure 9. Ethanol steam reforming on fresh (solid lines) and reactivated (dotted lines) catalysts. Ethanol conversion (**a**) NiCe P and (**b**) NiCe M and hydrogen yield (**c**) NiCe P and (**d**) NiCe M vs. reaction time.

As a matter of fact, the causes of catalysts deactivation were quite different for two samples. In fact, the catalyst prepared via microemulsions, despite the faster deactivation with time, seemed to be reactivated during the regeneration step. Therefore, its complete deactivation in the first run was probably due to a reversible coke deposition. SEM analyses performed on used NiCe M catalyst had shown that it was almost completely covered by a compact form of carbon, either polymeric or graphitic. Such coke can be oxidized during regeneration step. On the contrary, the catalyst prepared via precipitation is more stable over time, but it cannot be fully regenerated in air. For this used sample, SEM pictures have shown that only a small portion of metal was covered by coke deposits, and it could be easily removed during the oxidative treatment of regeneration. Hence, this is not the main cause of deactivation, and sintering of the active phase had probably occurred. As demonstrated by TPR technique, NiCe P presented a lower interaction between support and active phase, and this could have determined its easier sintering.

4. Conclusions

The effect of the preparation method, namely, precipitation and microemulsion synthesis, has been evaluated for nickel–ceria-based catalysts. As expected, the microemulsion approach allowed us to prepare materials with smaller NiO dimensions and a defined interaction between NiO and support stronger than in the catalysts with precipitated supports, as evidenced by TPR analyses. At the same time, the samples prepared via precipitation had higher surface areas, well-defined spherical particles, and higher crystallinity. Therefore, the preparation method strongly affected the structural and chemical properties of catalysts. In ethanol steam reforming, the catalysts prepared via precipitation showed higher catalytic activity and stability, while those prepared via microemulsion deactivated very fast. Similar trends were found for La-promoted supports. Nevertheless, after the oxidative regeneration treatment, the NiCe P catalyst did not fully regain its properties, while NiCe M was completely reactivated. This indicated that the reasons for catalyst deactivation should be quite different. For catalysts prepared by precipitation, the deactivation was mostly due to sintering of the nickel particles that were not strongly interacting with the support. On the contrary, strong interaction between active phase and support in NiCe M preserved the material from sintering. However, the lower surface area and the low degree of crystallinity led to a rapid deactivation of the material caused by coke deposition.

Author Contributions: Conceptualization, M.S. and V.C.C.; methodology, A.M.A.; formal analysis, C.P.; investigation, E.G.; data curation, F.M.; writing original draft preparation, review and editing, F.M.; supervision, M.S.; funding acquisition M.S. and V.C.C. All authors have read and agreed to the published version of the manuscript.

Funding: This research was funded by MINECO project CTQ2015-71823-R and MICINN project RTI2018-101604-B-I00 (Spain).

Acknowledgments: The technical help of M. Sanchez with SEM measurements is gratefully acknowledged.

Conflicts of Interest: The authors declare no conflict of interest.

References

1. IEA. The Future of Hydrogen, IEA, Paris. Available online: https://www.iea.org/reports/the-future-of-hydrogen (accessed on 12 November 2020).
2. Wang, J.; Chen, H.; Tian, Y.; Yao, M.; Li, Y. Thermodynamic analysis of hydrogen production for fuel cells from oxidative steam reforming of methanol. *Fuel* **2012**, *97*, 805–811. [CrossRef]
3. Ball, M.; Weeda, M. The hydrogen economy—Vision or reality? *Int. J. Hydrogen Energy* **2015**, *40*, 7903–7919. [CrossRef]
4. Navarro, R.M.; Peña, A.M.A.; Fierro, J.G. Hydrogen Production Reactions from Carbon Feedstocks: Fossil Fuels and Biomass. *Chem. Rev.* **2007**, *107*, 3952–3991. [CrossRef]
5. Abe, J.; Popoola, A.; Ajenifuja, E.; Popoola, O. Hydrogen energy, economy and storage: Review and recommendation. *Int. J. Hydrogen Energy* **2019**, *44*, 15072–15086. [CrossRef]
6. Gallucci, F.F.; Basile, A.A.; Tosti, S.; Iulianelli, A.; Drioli, E. Methanol and ethanol steam reforming in membrane reactors: An experimental study. *Int. J. Hydrogen Energy* **2007**, *32*, 1201–1210. [CrossRef]
7. Liu, Z.; Senanayake, S.D.; Rodriguez, J.A. Catalysts for the Steam Reforming of Ethanol and Other Alcohols. *Ethanol* **2019**, *574*, 133–158. [CrossRef]
8. Martinelli, M.; Watson, C.D.; Jacobs, G. Sodium doping of Pt/m-ZrO2 promotes C–C scission and decarboxylation during ethanol steam reforming. *Int. J. Hydrogen Energy* **2020**, *45*, 18490–18501. [CrossRef]
9. Mironova, E.; Lytkina, A.; Ermilova, M.; Efimov, M.N.; Zemtsov, L.; Orekhova, N.; Karpacheva, G.; Bondarenko, G.; Muraviev, D.; Yaroslavtsev, A.B. Ethanol and methanol steam reforming on transition metal catalysts supported on detonation synthesis nanodiamonds for hydrogen production. *Int. J. Hydrogen Energy* **2015**, *40*, 3557–3565. [CrossRef]
10. Mattos, L.V.; Jacobs, G.; Davis, B.H.; Noronha, F.B. Production of Hydrogen from Ethanol: Review of Reaction Mechanism and Catalyst Deactivation. *Chem. Rev.* **2012**, *112*, 4094–4123. [CrossRef]
11. Słowik, G.; Greluk, M.; Rotko, M.; Machocki, A. Evolution of the structure of unpromoted and potassium-promoted ceria-supported nickel catalysts in the steam reforming of ethanol. *Appl. Catal. B Environ.* **2018**, *221*, 490–509. [CrossRef]
12. Xu, W.; Liu, Z.; Johnston-Peck, A.C.; Senanayake, S.D.; Zhou, G.; Stacchiola, D.; Stach, E.A.; Rodriguez, J.A. Steam Reforming of Ethanol on Ni/CeO2: Reaction Pathway and Interaction between Ni and the CeO2 Support. *ACS Catal.* **2013**, *3*, 975–984. [CrossRef]
13. Rodrigues, T.S.; E Silva, F.A.; Candido, E.G.; Da Silva, A.G.M.; Geonmonond, R.D.S.; Camargo, P.H.C.; Linardi, M.; Fonseca, F. Ethanol steam reforming: Understanding changes in the activity and stability of Rh/MxOy catalysts as function of the support. *J. Mater. Sci.* **2019**, *54*, 11400–11416. [CrossRef]
14. Arslan, A.; Doğu, T. Effect of calcination/reduction temperature of Ni impregnated CeO2–ZrO2 catalysts on hydrogen yield and coke minimization in low temperature reforming of ethanol. *Int. J. Hydrogen Energy* **2016**, *41*, 16752–16761. [CrossRef]
15. Montero, C.; Remiro, A.; Benito, P.L.; Bilbao, J.; Gayubo, A.G. Optimum operating conditions in ethanol steam reforming over a Ni/La2O3-αAl2O3 catalyst in a fluidized bed reactor. *Fuel Process. Technol.* **2018**, *169*, 207–216. [CrossRef]
16. Di Michele, A.; Dell'Angelo, A.; Tripodi, A.; Bahadori, E.; Sànchez, F.; Motta, D.; Dimitratos, N.; Rossetti, I.; Ramis, G.; Sanchez, F. Steam reforming of ethanol over Ni/MgAl2O4 catalysts. *Int. J. Hydrogen Energy* **2019**, *44*, 952–964. [CrossRef]
17. Compagnoni, M.; Tripodi, A.; Di Michele, A.; Sassi, A.P.; Signoretto, M.; Rossetti, I. Low temperature ethanol steam reforming for process intensification: New Ni/MOx–ZrO2 active and stable catalysts prepared by flame spray pyrolysis. *Int. J. Hydrogen Energy* **2017**, *42*, 28193–28213. [CrossRef]
18. Wu, Z.; Li, M.; Overbury, S.H. On the structure dependence of CO oxidation over CeO2 nanocrystals with well-defined surface planes. *J. Catal.* **2012**, *285*, 61–73. [CrossRef]
19. Manzoli, M.; Avgouropoulos, G.; Tabakova, T.; Papavasiliou, J.; Ioannides, T.; Boccuzzi, M.M.A.F. Preferential CO oxidation in H2-rich gas mixtures over Au/doped ceria catalysts. *Catal. Today* **2008**, *138*, 239–243. [CrossRef]

20. Pizzolitto, C.; Menegazzo, F.; Ghedini, E.; Innocenti, G.; Di Michele, A.; Cruciani, G.; Cavani, F.; Signoretto, M. Increase of Ceria Redox Ability by Lanthanum Addition on Ni Based Catalysts for Hydrogen Production. *ACS Sustain. Chem. Eng.* **2018**, *6*, 13867–13876. [CrossRef]
21. Aneggi, E.; Boaro, M.; Colussi, S.; de Leitenburg, C.; Trovarelli, A. Ceria-Based Materials in Catalysis: Historical Perspective and Future Trends. In *Handbook on the Physics and Chemistry of Rare Earths*; Elsevier: Amsterdam, The Netherlands, 2016. [CrossRef]
22. Menegazzo, F.; Pizzolitto, C.; Ghedini, E.; Di Michele, A.; Cruciani, G.; Signoretto, M. Development of La Doped Ni/CeO$_2$ for CH$_4$/CO$_2$ Reforming. *C* **2018**, *4*, 60. [CrossRef]
23. Laguna, O.H.; Centeno, M.A.; Boutonnet, M.; Odriozola, J.A. Au-supported on Fe-doped ceria solids prepared in water-in-oil microemulsions: Catalysts for CO oxidation. *Catal. Today* **2016**, *278*, 140–149. [CrossRef]
24. Elias, J.S.; Risch, M.; Giordano, L.; Mansour, A.N.; Shao-Horn, Y. Structure, Bonding, and Catalytic Activity of Monodisperse, Transition-Metal-Substituted CeO$_2$ Nanoparticles. *J. Am. Chem. Soc.* **2014**, *136*, 17193–17200. [CrossRef] [PubMed]
25. Ferencz, Z.; Erdőhelyi, A.; Baán, K.; Oszkó, A.; Óvári, L.; Kónya, Z.; Papp, C.; Steinrück, H.-P.; Kiss, J. Effects of Support and Rh Additive on Co-Based Catalysts in the Ethanol Steam Reforming Reaction. *ACS Catal.* **2014**, *4*, 1205–1218. [CrossRef]
26. Eriksson, S.; Nylén, U.; Rojas, S.; Boutonnet, M. Preparation of catalysts from microemulsions and their applications in heterogeneous catalysis. *Appl. Catal. A Gen.* **2004**, *265*, 207–219. [CrossRef]
27. Brunauer, S.; Emmett, P.H.; Teller, E. Adsorption of Gases in Multimolecular Layers. *J. Am. Chem. Soc.* **2005**, *60*, 309–319. [CrossRef]
28. Barrett, E.P.; Joyner, L.G.; Halenda, P.P. The Determination of Pore Volume and Area Distributions in Porous Substances. I. Computations from Nitrogen Isotherms. *J. Am. Chem. Soc.* **1951**, *73*, 373–380. [CrossRef]
29. Rossetti, I.; Lasso, J.; Nichele, V.; Signoretto, M.; Finocchio, E.; Ramis, G.; Di Michele, A. Silica and zirconia supported catalysts for the low-temperature ethanol steam reforming. *Appl. Catal. B Environ.* **2014**, *150–151*, 257–267. [CrossRef]
30. Pinna, F. Supported metal catalysts preparation. *Catal. Today* **1998**, *41*, 129–137. [CrossRef]
31. Balbuena, P.B.; Gubbins, K.E. Theoretical interpretation of adsorption behavior of simple fluids in slit pores. *Langmuir* **1993**, *9*, 1801–1814. [CrossRef]
32. Thommes, M.; Kaneko, K.; Neimark, A.V.; Olivier, J.P.; Rodriguez-Reinoso, F.; Rouquerol, J.; Sing, K.S. Physisorption of gases, with special reference to the evaluation of surface area and pore size distribution (IUPAC Technical Report). *Pure Appl. Chem.* **2015**, *87*, 1051–1069. [CrossRef]
33. Manzoli, M.; Menegazzo, F.; Signoretto, M.; Cruciani, G.; Pinna, F. Effects of synthetic parameters on the catalytic performance of Au/CeO2 for furfural oxidative esterification. *J. Catal.* **2015**, *330*, 465–473. [CrossRef]
34. Nichele, V.; Signoretto, M.; Menegazzo, F.; Rossetti, I.; Cruciani, G. Hydrogen production by ethanol steam reforming: Effect of the synthesis parameters on the activity of Ni/TiO$_2$ catalysts. *Int. J. Hydrogen Energy* **2014**, *39*, 4252–4258. [CrossRef]
35. Menegazzo, F.; Burti, P.; Signoretto, M.; Manzoli, M.; Vankova, S.; Boccuzzi, F.; Pinna, F.; Strukul, G. Effect of the addition of Au in zirconia and ceria supported Pd catalysts for the direct synthesis of hydrogen peroxide. *J. Catal.* **2008**, *257*, 369–381. [CrossRef]
36. Pinton, N.; Vidal, M.; Signoretto, M.; Martínez-Arias, A.; Corberan, V.C. Ethanol steam reforming on nanostructured catalysts of Ni, Co and CeO$_2$: Influence of synthesis method on activity, deactivation and regenerability. *Catal. Today* **2017**, *296*, 135–143. [CrossRef]

Communication

Remediation of Lead and Nickel Contaminated Soil Using Nanoscale Zero-Valent Iron (nZVI) Particles Synthesized Using Green Leaves: First Results

Nimita Francy [1], Subramanian Shanthakumar [1], Fulvia Chiampo [2,*] and Yendaluru Raja Sekhar [3]

1. Department of Environmental and Water Resources Engineering, School of Civil Engineering, Vellore Institute of Technology (VIT), Vellore 632014, India; nimitafrancy@gmail.com (N.F.); shanthakumar.s@vit.ac.in (S.S.)
2. Department of Applied Science and Technology, Politecnico di Torino, Corso Duca degli Abruzzi 24, 10129 Torino, Italy
3. Department of Thermal and Energy Engineering, School of Mechanical Engineering, Vellore Institute of Technology (VIT), Vellore 632014, India; rajasekhar.y@vit.ac.in
* Correspondence: fulvia.chiampo@polito.it; Tel.: +39-011-090-4685

Received: 16 October 2020; Accepted: 12 November 2020; Published: 13 November 2020

Abstract: Nanoscale zero-valent iron (nZVI) particles have proved to be effective in the remediation of chlorinated compounds and heavy metals from contaminated soil. The present study aimed to analyze the performance of nanoparticles synthesized from low-cost biomass (green leaves) as chemical precursors, namely *Azadirachta indica* (neem) and *Mentha longifolia* (mint) leaves. These leaves were chosen because huge amounts of them are present in the region of Vellore. These nanoparticles were used to remove lead and nickel from contaminated soil. Characterization of nZVI particles was conducted using the Scanning Electron Microscope (SEM), Transmission Electron Microscope (TEM), and Brunauer–Emmett–Teller isotherm (BET) techniques. Remediation was performed on two different soil samples, polluted with lead or nickel at an initial metal concentration around 250 mg/kg of soil. The results revealed that after 30 days, the lead removal efficiency with 0.1 g of nZVI particles/kg of soil was 26.9% by particles synthesized using neem leaves and 62.3% by particles synthesized using mint leaves. Similarly, nickel removal efficiency with 0.1 g of particles/kg of soil was 33.2% and 50.6%, respectively, by particles using neem and mint leaves. When the nanoparticle concentration was doubled, Pb and Ni removal improved, with similar trends obtained at a lower dosage (0.1 g of particles/kg of soil). These first results evidenced that: (1) the nZVI particles synthesized using green leaves had the potential to remove Pb and Ni from contaminated soil; (2) the neem-derived particles gave better Ni removal efficiency than Pb one; (3) the mint-derived particles showed better Pb removal efficiency than Ni one; (4) the highest removal efficiency for both metals was achieved with the mint-derived particles; (5) double higher dosage did not greatly improve the results.

Keywords: neem; mint; nZVI synthesis; lead; nickel; soil remediation

1. Introduction

The discharge of industrial waste into the environment causes the accumulation of heavy metals in water and soil. Lead (Pb) and nickel (Ni) are two heavy metals present in the environment due to industrial activities. The average concentration of Pb on surface soil worldwide is approximately 32 mg/kg of soil [1], and its excessive amount is hazardous to living organisms and the environment [2]. Lead toxicity affects every part of the human body, mainly the nervous system and kidney. Children exposed to high Pb concentrations for a long time are likely to have impaired development [3].

The presence of Pb in high concentrations can affect joints of the human body and lead to miscarriage in pregnant women [4]. The primary sources of Pb discharge into the environment are mostly fertilizer, biosolids, metal mining, battery, and paint industries [5]. Generally, fertilizers used for supplying nitrogen–phosphorous–potassium (NPK) to soil contain lead and cadmium (Cd) as component elements [6]. Biosolids can contain Pb, depending on their industrial source. Pb concentration in soil equal to 300 µg/kg of soil is evaluated as a threshold without substantial risk for intake by humans [7].

In the metal mining industry, a large quantity of ores is extensively mined, causing potential risk to the environment by polluting the soil. Contamination of Ni to the environment is mostly from electroplating, welding, refining, battery, paint, and porcelain production industrial effluents [8–10]. Some of the significant health issues caused by Ni are nausea, gastric problems, and bronchitis [11]. The average concentration of Ni on the Earth's surface is 20 ppm [12], whereas the range of Ni concentration in soil is between 0.2 and 450 ppm [13].

The leaching of heavy metals due to rainfall causes the contamination of the surface, subsurface soil, and groundwater, affecting the food chain by bioaccumulation in each trophic level [14]. Numerous techniques have been adopted in the decontamination of heavy metals from soil, such as phytoremediation, phytoextraction, soil washing, adsorption, and nanoremediation [6]. Among the techniques, nanoremediation is an effective system as it contains smaller-sized active nanoparticles, with a large specific surface area [15]. Among these, nanoscale zero-valent iron (nZVI) particles are nanoparticles (1–100 nm) containing zero-valent iron, obtained from different kinds of chemical synthesis. As for other pollutants, the mechanisms for the removal of heavy metals by nZVI particles are adsorption, reduction/oxidation, precipitation, and coprecipitation.

Pasinszki and Krebsz [16] recently published a review on the synthesis and application of nZVI nanoparticles. This review constitutes one of the most comprehensive and prominent reviews on the use of these nanoparticles, completed with an interesting section on the particle toxicity.

The type and nature of the remediation process undertaken by nZVI mainly depend on the electronegativity of the contaminants to be removed [17]. Literature data show that: (a) nZVI with 1% hydrogen peroxide is effective for the decontamination of pentachlorophenol [18]; (b) MgO, TiO_2, and ZnO nanoparticles at 1% proved to provide good removal of chromium from soil contaminated with leather factory waste [19]; (c) the contact particle air is deleterious due to oxide formation, and preparation of stable nanoparticles was proposed with different chelating agents [20]. Hence, nanoremediation using nZVI particles is an emerging technique for wastewater decontamination and soil remediation, with more questions remaining.

Regarding water/wastewater treatment, nanoparticle application has received wider attention and development than soil remediation due to the easiness of contact water particles.

Valipour et al. [21] conducted studies to evaluate remediation characteristics of two phosphorus amendments, triple superphosphate (TSP) and phosphate rock (PR), to reduce Pb, Cd, Ni, and Cu contamination in four artificially contaminated, mainly calcareous, soils. Though TSP reduced the Pb and Cd presence, it increased the availability of Ni. PR did not show any reduction of metal contamination in calcareous soils. Yadegari [22] studied the influence on growing purslane plants to reduce the contamination of heavy metals such as Ni and Cd. He conducted pot experiments by spiking Ni (0, 30, 60, and 120 mg/kg of soil) and Cd (0, 10, 20, and 40 mg/kg of soil) into soil for two seasons. Heavy metals in the soil had a compelling effect on the morphological and physiological characteristics of purslane. Higher concentrations of metal contamination resulted in a decrease of morphological and physiological characteristics and a stronger influence of Cd. De Gisi et al. [23] used commercially available nZVI Nanofer 25S to treat contaminated marine sediments polluted by heavy metals. They conducted experimental runs on soil particles <5 mm and two dosages, i.e., low dosage (2, 3, and 4 g nZVI per kg of Suspended Solids) and high dosage (5, 10, and 20 g nZVI per kg of Suspended Solids). They concluded that the average dosages of nZVI could effectively reduce heavy metal contamination in sediments. Vasarevičius et al. [24] conducted experimental runs to remove Cd, Cu, Ni, and Pb contamination in spiked soil samples using commercial nZVI particles. They evaluated the remediation levels for single and multiple metals (mixtures of Cu, Ni, Pb and Cd, Cu, Ni, Pb) using different doses (0%, 0.85%, 1.7%, 2.55%, and 5.1% by weight) of nZVI particles. The leaching procedure was adopted to determine immobilization efficiency for each specific metal and nZVI dose. Their results showed effective metal removal and metal stabilization at higher dosages for all the samples.

In the present study, nanoscale zero-valent iron (nZVI) particles were chemically synthesized using neem and mint leaves to remove lead and nickel from two soils. These leaves were chosen because huge amounts of them are present in the region of Vellore.

The preliminary results are promising and worth future studies to better understand their performance and find optimal conditions for their application at a larger scale.

2. Materials and Methods

2.1. Synthesis of nZVI Particles

Synthesis of nZVI particles using $FeSO_4 \cdot 7H_2O$ and $NaBH_4$: For nZVI synthesis, 4.17 g of $FeSO_4 \cdot 7H_2O$ in 150 mL of water and 3.72 g of EDTA in 100 mL of water were mixed with 2.87 g of $NaBH_4$ in 100 mL of distilled water in a stirred closed flask for 60 min and purged with nitrogen to achieve black color precipitates. These particles were filtered, washed with ethanol thrice, and filtered again [20].

Synthesis of nZVI particles using neem and mint leaves: Synthesis using *Azadirachta indica* (neem) and *Mentha longifolia* (mint) leaves was identical for each kind of leaf and involved the following procedure: (i) 10 g of leaves were cleaned, washed, dried, chopped and grounded; (ii) the leaves were then heated to 50 °C within 20 mL of distilled water for 10 min; (iii) 20 mL of each leaf extract was mixed with 20 mL of 0.001 M $FeCl_3 \cdot 6H_2O$ in a stirred closed flask for 60 min and purged with nitrogen. During this stage, 1% of polyvinyl alcohol and 1% of chitosan were added to stabilize the particles; (iv) the obtained black color particles were filtered and (v) dried at 50 °C for 12 h [25].

2.2. Characterization of nZVI Particles

Each kind of nanoparticles was studied using the Scanning Electron Microscope (SEM), Transmission Electron Microscope (TEM), and Brunauer–Emmett–Teller isotherm (BET). SEM analysis was performed to examine the surface morphology, whereas the shape and the approximate structure could be achieved from TEM analysis. BET analysis was useful to determine the specific surface area.

2.3. Soil and its Original Characteristics

Two soil samples (A and B) of 1 kg each were collected from different locations in VIT University, Vellore. The samples were cleaned, dried, and sieved through a 2 mm sieve. The pH and electrical conductivity of each soil sample were determined as per the Bureau of Indian Standards [26,27]. Specific gravity [28] (ASTM-D854, 2014), water content [29] (ASTM-D4959, 2016), organic content [30] (ASTM-D2974, 2014), and particle size analysis [31] (ASTM-D422, 2007) of the soil samples were conducted.

The initial lead and nickel content of the soil was determined by the Atomic Absorption Spectroscopy (AAS) technique. The soil extract was prepared with 5 g of soil in 50 mL of 1 M HNO_3 and placed in the shaker for 60 min to ascertain complete mixing. The supernatant obtained was used for AAS analysis.

The same method was used to monitor the experimental runs to measure the residual concentration of lead or nickel in the soil.

2.4. Pollution of Soil Samples and Nanoparticle Addition

The heavy-metal-contaminated soil was prepared by adding Ni or Pb solution to the soil. For nickel, a solution was prepared by dissolving 4.785 g of $NiSO_4 \cdot 7H_2O$ in 1 L of double-distilled water. Similarly, the lead solution was made by dissolving 1.615 g of Pb(NO$_3$)$_2$ in 1 L of double-distilled water. Spiking of the soil was conducted to achieve a target metal concentration around 250 mg/kg of soil and, at the same time, considering the solution water content as a contribution to meet the soil field capacity. The initial heavy metal concentration was constantly verified by analysis.

The addition of nanoparticles was conducted on parts of the initial sample to optimize the distribution. Each soil sample and nanoparticle dose were divided into five parts. Then, each soil part was mixed with one particle part. Finally, these 5 quotas were transferred to a 1 L beaker and monitored for 30 days. Two dosages were tested, namely 0.1 and 0.2 g/kg of soil.

2.5. Remediation of Soil Samples

The remediation of heavy-metal-contaminated soil was studied at room temperature for 30 days. Along each test period, Pb and Ni residual concentration was monitored by AAS analysis on the extract derived from the 5 g sample leaching with HNO$_3$ (see Section 2.3), and the effect of concentration and time on soil was studied.

All the analyses were replicated twice. The average results will be shown without error bars and standard deviations because two replicates cannot support a reliable statistical analysis.

The residual concentration at time t = 30 days was used to calculate the removal efficiency:

$$\text{Heavy metal removal \%} = \frac{C_i - C_f}{C_i} \times 100 \tag{1}$$

where: C_i = initial heavy metal concentration (mg/kg of soil), C_f = final heavy metal concentration (mg/kg of soil).

3. Results

3.1. Characteristics of nZVI Particles

BET analysis: The surface area of the chemically synthesized nZVI particles was 15.2 m^2/g, and for the particles synthesized using neem and mint leaves, the surface area was 6.2 m^2/g and 13.0 m^2/g, respectively.

SEM analysis: The SEM images are presented in Figure 1a,b (chemically synthesized nanoparticles), Figure 1c,d (nanoparticles synthesized using neem leaves), and Figure 1e,f (nanoparticles synthesized using mint leaves). The images show that the particles are spherical and subjected to agglomeration.

TEM analysis: The TEM images are shown in Figure 2. It can be observed that the particles derived from leaf extracts are agglomerated. The color difference between the core and outer layer of the nanoparticle shows that the chemically synthesized nanoparticle (Figure 2a,b) is more subjected to oxidation than those synthesized using leaves (Figure 2c–f).

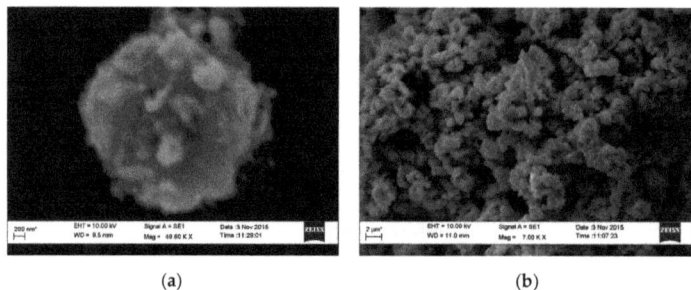

(a)   (b)

Figure 1. *Cont.*

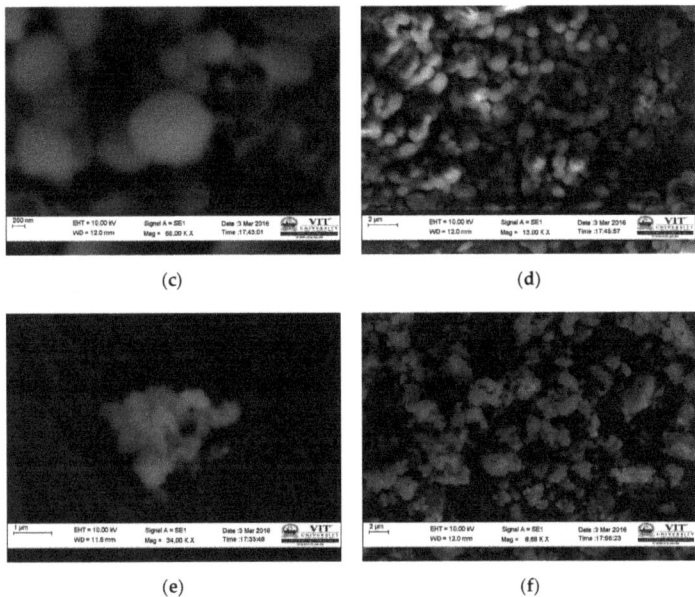

Figure 1. SEM image of nanoscale zero-valent iron (nZVI) particles: (**a**,**b**) from chemical synthesis; (**c**,**d**) using neem leaves; (**e**,**f**) using mint leaves.

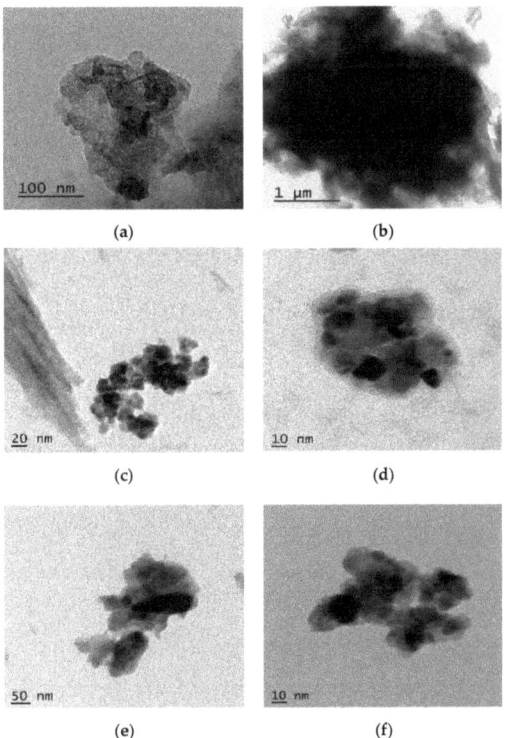

Figure 2. TEM image of nZVI particles: (**a**,**b**) chemically synthesized; (**c**,**d**) using neem leaves; (**e**,**f**) using mint leaves.

3.2. Soil Characteristics

The initial characteristics of the collected soil samples are presented in Table 1. Both soils were categorized as coarse-graded, sandy soil and contained a trace of lead and nickel, as demonstrated by their low concentration.

Table 1. Initial characteristics of the soil samples.

Characteristics	Soil A	Soil B
Soil type	Coarse-graded, sandy soil	Coarse-graded, sandy soil
pH	6.65	5.32
Conductivity (mS/cm)	1.12	0.98
Water content (%)	13.7	12.9
Specific gravity	2.98	3.07
Organic content (% by weight)	2.78	4.94
Lead concentration (mg/kg of soil)	0.245	0.234
Nickel concentration (mg/kg of soil)	0.201	0.267

3.3. Remediation of Contaminated Soil

3.3.1. Remediation by Chemically Synthesized nZVI Particles

The remediation monitoring of lead and nickel using chemically synthesized nZVI particles is presented in Figure 3, where the residual lead and nickel concentration at the dosage of 0.1 g of nanoparticle/kg of soil added to each contaminated soil sample (A and B) is presented.

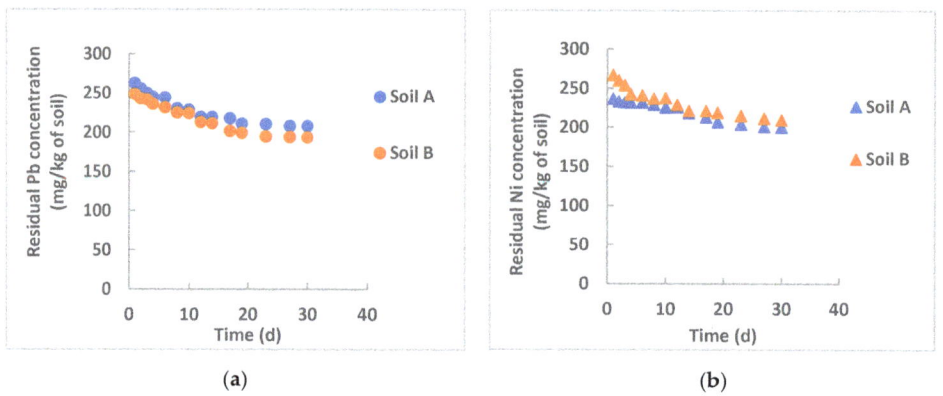

Figure 3. Remediation of contaminated soil using chemically synthesized nZVI particles (dosage: 0.1 g/kg of soil): (**a**) lead removal; (**b**) nickel removal.

Considering the system heterogeneity, the experimental data show similar values for the tested soil samples, and on this basis, the heavy metal removal efficiency was calculated as the average value at the end of the experimental run.

After 30 days and with a particle dosage of 0.1 g/kg of soil, the Pb removal efficiency was 21.6% and 18.5% for nickel. It can be observed that in any instance, the contaminant removal efficiency of the nanoparticle is low. This finding could be due to aggregation and/or oxidation of nanoparticles. In general, oxidation of the outer layer of the nanoparticle is mainly due to contact with air, which shrinks the adsorption capability of the nanoparticle. This hypothesis is supported by the SEM and TEM analyses. The SEM images show that the nanoparticles are aggregated, which reduces the surface area of the particle. The TEM images indicate the oxidation of the outer layer of nanoparticles.

3.3.2. Remediation by nZVI Particles Synthesized Using Neem Leaves

The nZVI particles synthesized using neem leaves were tested in the same conditions as the chemically synthesized particles, and the experimental results are shown in Figure 4.

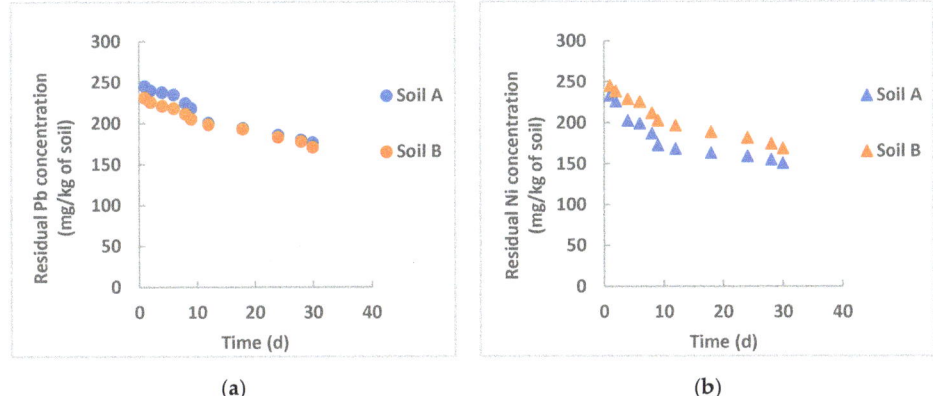

Figure 4. Remediation of nZVI particles synthesized using neem leaves (dosage: 0.1 g/kg of soil): (**a**) lead removal; (**b**) nickel removal.

Figure 4a,b shows the residual lead and nickel concentration, respectively, along with the tests with Soils A and B at a dosage of 0.1 g of neem synthesized nZVI/kg of soil.

Regarding the chemically synthesized particles, Soils A and B performed similarly. After 30 days, the removal efficiency of lead at a dosage of 0.1 g of neem synthesized nZVI/kg of soil was 26.9%, and for nickel, it was 33.2%, demonstrating better performance with nickel than with lead.

After this, it was decided to test double particle dosage with the same initial heavy metal concentration.

Figure 5 depicts the monitoring with 0.2 g of nZVI particles synthesized using neem leaves per kg of soil.

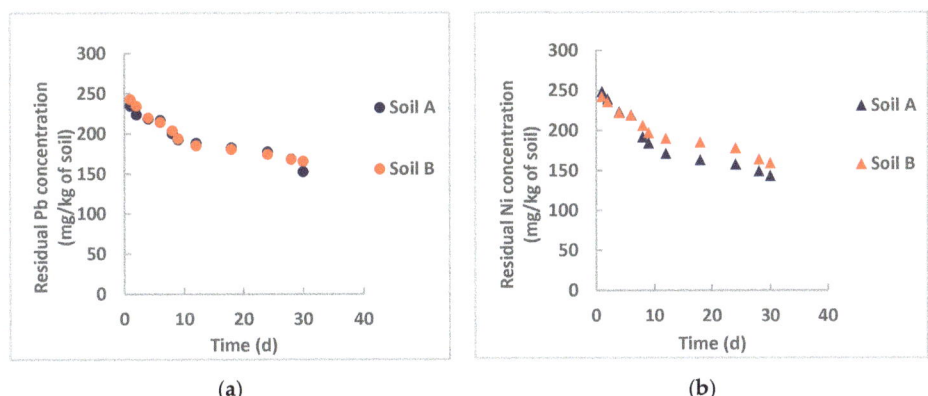

Figure 5. Remediation of nZVI particles synthesized using neem leaves (dosage: 0.2 g/kg of soil): (**a**) lead removal; (**b**) nickel removal.

The trend is the same as that achieved with a dosage of 0.1 g/kg of soil. The lead and nickel removal efficiency was observed as 33.3% and 38.2%, respectively. Comparing these values to lower values, it is evident that the improvement is limited in the order of 24% for lead and 15% for nickel.

Examining all the findings, the removal efficiency is not high. This could be due to the low surface area measured by the BET analysis. Agglomeration and oxidation of nanoparticles play a vital role in the reduction of removal efficiencies, and unfortunately, these phenomena are evidenced by SEM and TEM images, respectively.

3.3.3. Remediation by nZVI Particles Synthesized Using Mint Leaves

The particles derived from mint leaves were tested at the same dosage previously used, namely 0.1 g/kg of soil. Figure 6 reports the monitoring during the 30-day test.

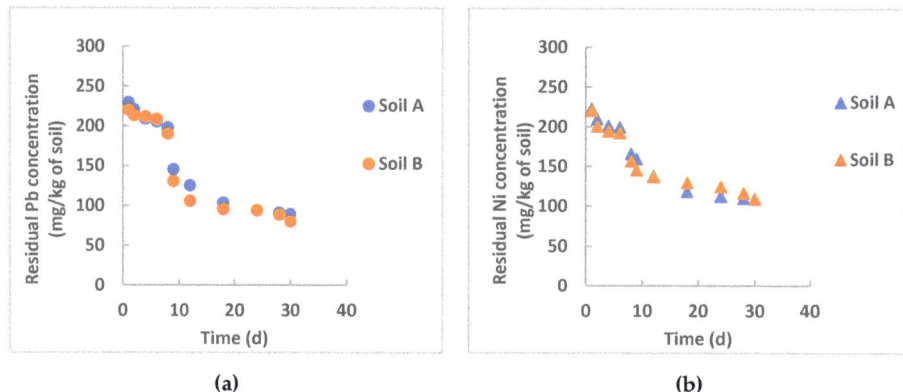

Figure 6. Remediation of nZVI particles synthesized using mint leaves (dosage: 0.1 g/kg of soil): (**a**) lead removal; (**b**) nickel removal.

The tested soils behaved similarly, probably due to their similar properties. At the end of the test (after 30 days), the removal efficiency was 62.3% (lead removal) and 50.6% (nickel removal).

In this instance, the particles showed better performance with lead than nickel. In both instances, the values were much higher than those with particles derived using neem leaves.

Figure 7 shows the removal efficiency at a doubled dosage of particles derived using mint leaves (0.2 g/kg of soil).

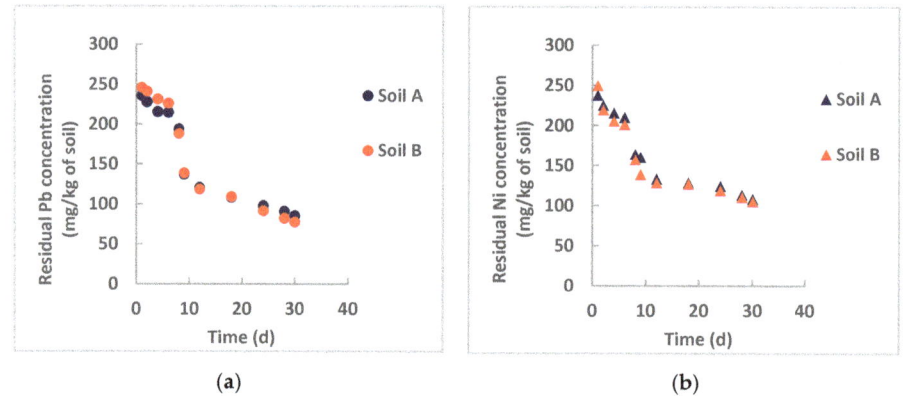

Figure 7. Remediation of nZVI particles synthesized using mint leaves (dosage: 0.2 g/kg of soil): (**a**) lead removal; (**b**) nickel removal.

The experimental data do not show relevant differences between the tested soils.

The removal efficiency was higher for lead than nickel, confirming the trend achieved with a lower dosage. After 30 days, 66.1% of the initial lead and 56.1% of the initial nickel was removed.

One reason for the better performance of particles synthesized using mint leaves could be the agglomeration, which is comparatively less than for neem-derived particles, and hence, their surface area is higher than for neem-derived particles (Figure 1).

When the particles were oxidized on their outer surface, the oxidation rate was lower than that of the chemically synthesized particles, as can be observed from the TEM images (Figure 2).

4. Discussion

This study aimed to provide preliminary results about the performance of nZVI particles derived using green leaves to remediate heavy-metal-polluted soil. At the same time, chemically synthesized nanoparticles were prepared to compare their removal efficiency to those achievable by vegetal-origin nanoparticles. As leaves, *Azadirachta indica* (neem) and *Mentha longifolia* (mint) were chosen.

The targeted heavy metals were lead and nickel.

For a rational discussion and comparison easiness, the results for removal efficiency are summarized in Table 2.

Table 2. The heavy metal removal efficiency of the tested nZVI particles.

Particle Origin	Pb Removal Efficiency at t = 30 days		Ni Removal Efficiency at t = 30 days	
	Particle dosage			
	0.1 g/kg of soil	0.2 g/kg of soil	0.1 g/kg of soil	0.2 g/kg of soil
Chemically synthesized	21.6%		18.5%	
Neem leaves	26.9%	33.3%	33.2%	38.2%
Mint leaves	62.3%	66.1%	50.6%	56.1%

By these values, some conclusions can be evidenced:

- The chemically synthesized particles provided the lowest efficiencies.
- The particles achieved by the processing of mint leaves showed the best results for the removal of both the metals.
- With identical nanoparticles dosage, the removal of lead by all the tested particles was constantly higher than nickel, suggesting a higher affinity of the particle for the metal.
- A double dosage improved the removal, albeit to a small extent (the maximum improvement was 26% for lead removal by neem-derived particles).

To summarize, the ZVI nanoparticles synthesized using green leaves demonstrated a good performance to remove lead and nickel from polluted soil, better than those obtained from chemical synthesis.

To this purpose, different particle properties could play opposite roles. The chemically synthesized particles showed higher BET surface area of approximately 15 against 6–13 m^2/g of the others. Therefore, the adsorption should occur to a larger extent. However, as a consequence of this property, they were also highly subjected to oxidation, as shown by the TEM images, reducing their removal efficiency.

Examining the residual heavy metal concentration, when chemically synthesized particles were adopted, a decreasing trend was continuous, whereas in the other tests, two-step process could be identified: the first one, lasting about 8–10 days, with a drastic decrease of residual concentration, followed by further removal at a slower rate until the end of the test.

A similar trend was found by other authors:

- Gil-Diaz et al. [32] studied in-field brownfield remediation polluted with arsenic and mercury using commercial nZVI particles for long times (32 months) and achieved an initial sharp reduction of the pollutants, followed by a rather constant residual concentration probably due to interferences of organic matter that can form stable complexes on the particle surface, especially at acidic pH and limiting the particle removal efficiency. Examining Figures 3–5, this occurs for nickel removal in Soil B, where pH is moderately acidic (5.32) and the organic matter content is high (about 5%). This is evident for nickel, whereas the pH influence is not appreciable for lead. One hypothesis can be the different solubility product constant, K_{PS}, for $Ni(OH)_2$ and $Pb(OH)_2$, which is equal to 6×10^{-16} and 1.4×10^{-20}, respectively. A rough calculation can provide the saturation concentration of the heavy metals at the tested pH values. For lead, at saturation its concentration is lower than the initial concentration in both instances. Therefore, lead hydroxide also precipitates at an acidic pH. Heavy metal removal by nZVI particles is not only based on precipitation. For lead, this phenomenon could be more influential than the others (adsorption, coprecipitation, oxidation/reduction);

- Mystrioti et al. [33] studied the reduction of hexavalent chromium to Cr(III) by nZVI particles synthesized from several sources, namely *Camellia sinensis* (green tea), *Syzygium aromaticum* (clove), *Mentha spicata* (spearmint), *Punica granatum juice* (pomegranate), and red wine. The process was conducted on a liquid solution containing 50 mg/L of Cr(VI) in contact with different nanoparticle concentrations from the aforementioned sources. The reduction process, occurring on the particle surfaces, had better efficiency at a high particle concentration, and when a low concentration was used, the process kinetics clearly showed two different rates.
- Di Palma et al. [34] used chemically synthesized nZVI particles to reduce Cr(VI) to Cr(III) from neutral soil in the slurry mode. Their trials also showed that the reduction efficiency was influenced positively by the test duration and nZVI particle concentration, with a more evident two-step process at a low nZVI particle concentration.
- Wang et al. [35] synthesized nZVI particles from green tea and eucalyptus leaves to remove nitrate from wastewater and compared their performance to the results achieved with particles from chemical synthesis. The best results were achieved with the nanoparticles of chemical origin. However, after air contact for two months, the vegetal-origin nanoparticles did not change their performance and showed good stability, whereas for the others, the removal efficiency dropped by 50%.

From this, it is evident that the findings of the present study are widely supported by similar studies, even if performed with different operative conditions of pollutants, particle origin, and concentration, soil properties, etc.

These findings also evidence the need to continue the study and clarify the many features still pending to apply the process optimally and achieve high efficiency in sustainable terms.

Author Contributions: Conceptualization, S.S.; methodology, S.S. and Y.R.S.; validation, S.S.; formal analysis, S.S. and F.C.; investigation, N.F.; resources, S.S.; writing—original draft preparation, S.S.; writing—review and editing, S.S. and F.C.; supervision, S.S. All authors have read and agreed to the published version of the manuscript.

Funding: This research received no external funding.

Acknowledgments: The first author (N.F.) acknowledges the financial support provided by The Institution of Engineers (India), Kolkata, India under the R & D Grant-in-aid Scheme (Grant No. RDPG2016004).

Conflicts of Interest: The authors declare no conflict of interest.

References

1. Kabata-Pendias, A.; Pendias, H. *Trace Metals in Soils and Plants*; CRC Press: Boca Raton, FL, USA, 2001.
2. Usha, R.; Vasavi, A.; Thishya, K.; Janshi Rani, S.; Supraja, P. Phytoextraction of lead from industrial effluents by sunflower (*Helianthus Annuus. L*). *Rasayan J. Chem.* **2011**, *4*, 8–12.
3. Burke, D.M.; Morris, M.A.; Holmes, J.D. Chemical oxidation of mesoporous carbon foams for lead ion adsorption. *Sep. Purif. Technol.* **2013**, *104*, 150–159. [CrossRef]
4. National Safety Council. *Lead Poisoning*. Available online: https://www.nsc.org/new_resources/%20resources/document/lead_poisoning (accessed on 18 November 2015).

5. Özcan, A.S.; Tunali, S.; Akar, T.; Özcan, A. Biosorption of Lead (II) ions onto waste biomass of *Phaseolus vulgaris* L.: Estimation of the equilibrium, kinetic and thermodynamic parameters. *Desalination* **2009**, *244*, 188–198. [CrossRef]
6. Wuana, R.A.; Okieimen, F.E. Heavy metals in contaminated soils: A review of sources, chemistry, risk and best, available strategies for remediation. *ISRN Ecol.* **2011**. ID 402647. [CrossRef]
7. Dudka, S.; Miller, W.P. Permissible concentration of arsenic and lead in soils based on risk assessment. *Water Air Soil Pollut.* **1999**, *113*, 127–132. [CrossRef]
8. Akhtar, N.; Iqbal, J.; Iqbal, M. Removal and recovery of nickel (II) from aqueous solution by loofa sponge-immobilized biomass of *Chlorella sorakiniana*: Characterisation study. *J. Hazard. Mater.* **2004**, *B108*, 85–94. [CrossRef]
9. Farooq, U.; Kozinski, J.A.; Khan, M.A.; Athar, M. Biosorption of heavy metal ions using wheat-based biosorbents: A review of the recent literature. *Bioresour. Technol.* **2010**, *101*, 5034–5043. [CrossRef]
10. Senthil Kumar, P.S.; Ramalingam, S.; Dinesh Kirupha, S.; Murugesan, A.; Vidhyadevi, T.; Sivanesan, S. Adsorption behaviour of nickel (II) onto cashew nut shell: Equilibrium thermodynamics, kinetics, mechanism and process design. *Chem. Eng. J.* **2011**, *167*, 122–131. [CrossRef]
11. Malamis, S.; Katsou, E. A review on zinc and nickel adsorption on natural and modified zeolite, bentonite and vermiculite: Examination of process parameters, kinetics and isotherms. *J. Hazard. Mater.* **2013**, *252–253*, 428–461. [CrossRef]
12. Adriano, D.C. *Trace Elements in Terrestrial Environment*; Springer: New York, NY, USA, 2001.
13. Iyaka, Y.A. Nickel in soils: A review of its distribution and impacts. *Sci. Res. Essays* **2011**, *6*, 6774–6777.
14. Basta, N.T.; Gradwohl, R. Remediation of heavy metals contaminated soil using rock phosphate. *Better Crops* **1998**, *82*, 29–31.
15. Kuiken, T. Cleaning up contaminated waste site: Is nanotechnology the answer? *Nano Today* **2009**, *5*, 6–8. [CrossRef]
16. Pasinszki, T.; Krebsz, M. Synthesis and Application of Zero-Valent Iron Nanoparticles in Water Treatment, Environmental Remediation, Catalysis, and Their Biological Effects. *Nanomaterials* **2020**, *10*, 917. [CrossRef]
17. Bhattacharya, S.; Saha, I.; Mukhopadhyay, A.; Chattopadhyay, D.; Gosh, U.; Chatterjee, D. Role of nanotechnology in water treatment and purification: Potential applications and implications. *Int. J. Chem. Sci. Technol.* **2013**, *3*, 59–64.
18. Liao, C.J.; Chung, T.L.; Chen, W.L.; Kuo, S.L. Treatment of pentachlorophenol-contaminated soil using nano-scale zero-valent iron with hydrogen peroxide. *J. Mol. Catal. A Chem.* **2006**, *265*, 189–194. [CrossRef]
19. Taghipour, M.; Jalali, M. Effect of clay minerals and nanoparticles on chromium fractionation in soil contaminated with leather factory waste. *J. Hazard. Mater.* **2015**, *297*, 127–133. [CrossRef]
20. Allabaksh, M.B.; Mandal, B.K.; Kesarla, M.K.; Kumar, K.S.; Reddy, P.S. Preparation of stable Zero Valent Iron nanoparticles using different chelating agents. *J. Chem. Pharm. Res.* **2010**, *2*, 67–74.
21. Valipour, M.; Shahbazi, K.; Khanmirzaei, A. Chemical immobilization of lead, cadmium, copper, and nickel in contaminated soils by phosphate amendments. *CLEAN Soil Air Water* **2016**, *44*, 572–578. [CrossRef]
22. Yadegari, M. Performance of purslane (*Portulaca oleracea*) in nickel and cadmium contaminated soil as a heavy metals-removing crop. *Iran. J. Plant Physiol.* **2018**, *8*, 2447–2455.
23. De Gisi, S.; Minetto, D.; Lofrano, G.; Libralato, G.; Conte, B.; Todaro, F.; Notarnicola, M. Nano-scale zero valent iron (nZVI) treatment of marine sediments slightly polluted by heavy metals. *Chem. Eng. Trans.* **2017**, *60*, 139–144.
24. Vasarevičius, S.; Danila, V.; Paliulis, D. Application of stabilized nano Zero Valent Iron particles for immobilization of available Cd^{2+}, Cu^{2+}, Ni^{2+}, and Pb^{2+} ions in soil. *Int. J. Environ. Res.* **2019**, *13*, 465–474. [CrossRef]
25. Pattanayak, M.; Nayak, P.L. Green synthesis and characterisation of Zero Valent Iron nanoparticles from leaf extract of *Azadirachta Indica* (Neem). *World J. Nano Sci. Technol.* **2013**, *2*, 6–9.
26. *IS:3025-Part 11, Methods of Sampling and Test (Physical and Chemical) for Water and Wastewater, Part 11 pH Value*; Bureau of Indian Standards: New Delhi, India, 2006.
27. *IS:3025-Part 14, Methods of Sampling and Test (Physical and Chemical) for Water and Wastewater, Part 14 Specific Conductance*; Bureau of Indian Standards: New Delhi, India, 2002.
28. *ASTM D854, Standard Test Methods for Specific Gravity of Soil Solids by Water Pycnometer*; ASTM International: West Conshohocken, PA, USA, 2014.

29. *ASTM D4959, Standard Test Method for Determination of Water Content of Soil by Direct Heating*; ASTM International: West Conshohocken, PA, USA, 2016.
30. *ASTM D2974, Standard Test Methods for Moisture, Ash, and Organic Matter of Peat and Other Organic Soils*; ASTM International: West Conshohocken, PA, USA, 2014.
31. *ASTM D422, Standard Test Method for Particle-Size Analysis of Soils*; ASTM International: West Conshohocken, PA, USA, 2007.
32. Gil-Díaz, M.; Rodríguez-Valdés, E.; Alonso, J.; Baragaño, D.; Gallego, J.R.; Lobo, M.C. Nanoremediation and long-term monitoring of brownfield soil highly polluted with As and Hg. *Sci. Total Environ.* **2019**, *675*, 165–175. [CrossRef]
33. Mystrioti, C.; Xanthopoulou, T.D.; Tsakiridis, P.; Papassiopi, N.; Xenidis, A. Comparative evaluation of five plant extracts and juices for nanoiron synthesis and application for hexavalent chromium reduction. *Sci. Total Environ.* **2016**, *539*, 105–113. [CrossRef]
34. Di Palma, L.; Gueye, M.T.; Petrucci, E. Hexavalent chromium reduction in contaminated soil: A comparison between ferrous sulphate and nanoscale zero-valent iron. *J. Hazard. Mater.* **2015**, *281*, 70–76. [CrossRef]
35. Wang, T.; Lin, J.; Chen, Z.; Megharaj, M.; Naidu, R. Green synthesized iron nanoparticles by green tea and eucalyptus leaves extracts used for removal of nitrate in aqueous solution. *J. Clean. Prod.* **2014**, *83*, 413–419. [CrossRef]

Publisher's Note: MDPI stays neutral with regard to jurisdictional claims in published maps and institutional affiliations.

© 2020 by the authors. Licensee MDPI, Basel, Switzerland. This article is an open access article distributed under the terms and conditions of the Creative Commons Attribution (CC BY) license (http://creativecommons.org/licenses/by/4.0/).

Article

Highly Selective Syngas/H_2 Production via Partial Oxidation of CH_4 Using (Ni, Co and Ni–Co)/ZrO_2–Al_2O_3 Catalysts: Influence of Calcination Temperature

Anis Hamza Fakeeha [1], Yasir Arafat [1], Ahmed Aidid Ibrahim [1], Hamid Shaikh [2], Hanan Atia [3,*], Ahmed Elhag Abasaeed [1], Udo Armbruster [3] and Ahmed Sadeq Al-Fatesh [1,*]

1. Chemical Engineering Department, College of Engineering, King Saud University, P.O. Box 800, Riyadh 11421, Saudi Arabia; anishf@ksu.edu.sa (A.H.F.); engr.arafat111@yahoo.com (Y.A.); aidid@ksu.edu.sa (A.A.I.); abasaeed@ksu.edu.sa (A.E.A.)
2. Chemical Engineering Department, SABIC Polymer Research Center, King Saud University, P.O. Box 800, Riyadh 11421, Saudi Arabia; hamshaikh@ksu.edu.sa
3. Leibniz Institute for Catalysis, 18055 Rostock, Germany; udo.armbruster@catalysis.de
* Correspondence: hanan.atia@catalysis.de (H.A.); aalfatesh@ksu.edu.sa (A.S.A.-F.); Tel.: +49-3811281258 (H.A.); +966-14676859 (A.S.A.-F.)

Received: 7 February 2019; Accepted: 28 February 2019; Published: 6 March 2019

Abstract: In this study, Ni, Co and Ni–Co catalysts supported on binary oxide ZrO_2–Al_2O_3 were synthesized by sol-gel method and characterized by means of various analytical techniques such as XRD, BET, TPR, TPD, TGA, SEM, and TEM. This catalytic system was then tested for syngas respective H_2 production via partial oxidation of methane at 700 °C and 800 °C. The influence of calcination temperatures was studied and their impact on catalytic activity and stability was evaluated. It was observed that increasing the calcination temperature from 550 °C to 800 °C and addition of ZrO_2 to Al_2O_3 enhances Ni metal-support interaction. This increases the catalytic activity and sintering resistance. Furthermore, ZrO_2 provides higher oxygen storage capacity and stronger Lewis basicity which contributed to coke suppression, eventually leading to a more stable catalyst. It was also observed that, contrary to bimetallic catalysts, monometallic catalysts exhibit higher activity with higher calcination temperature. At the same time, Co and Ni–Co-based catalysts exhibit higher activity than Ni-based catalysts which was not expected. The Co-based catalyst calcined at 800 °C demonstrated excellent stability over 24 h on stream. In general, all catalysts demonstrated high CH_4 conversion and exceptionally high selectivity to H_2 (~98%) at 700 °C.

Keywords: Al_2O_3; bimetallic catalyst; syngas; methane; partial oxidation; ZrO_2

1. Introduction

Methane (CH_4) is an important constituent of natural and biogas and plays an important role in C_1 chemistry. Its utilization is expected to increase in the future because of the weaker greenhouse gas effect (CO_2 release) compared to other fossil resources. However, it is well known that the direct conversion of methane yields less valuable petrochemical products and hence it is necessary to resort to an indirect process that initially involves the generation of synthesis gas (H_2 and CO) [1–4]. Synthesis gas is widely used in the production of hydrogen, synthetic fuels, alcohols and other chemicals. It can be produced by partial oxidation of hydrocarbons, particularly methane, via (i) steam reforming or (ii) dry reforming (DRM) or (iii) autothermal reforming. Specifically, the catalytic partial oxidation of methane has been recognized as a beneficial process from both technical and economic perspective; as it requires less energy and capital cost due to low endothermic nature of the process [5]. In addition, the H_2/CO ratio of 2 is suitable for methanol synthesis and higher hydrocarbons through the Fischer-Tropsch process [6].

Various reaction mechanisms have been suggested for the partial oxidation of methane. The first is a direct route (Equation (1)) while the second mechanism comprises combustion and two reforming reactions. In the latter pathway, combustion of methane is accomplished (Equation (2)). Subsequently, steam and dry reforming of methane take place in the presence of the newly produced CO_2 and H_2O, respectively (Equations (3) and (4)) to render syngas.

$$CH_4 + 0.5O_2 \rightarrow CO + 2H_2 \qquad \Delta H^{\circ}_{298k} = -35.7 \text{ kJ/mol} \qquad (1)$$

$$CH_4 + 2O_2 \rightarrow CO_2 + 2H_2O \qquad \Delta H^{\circ}_{298k} = -802.3 \text{ kJ/mol} \qquad (2)$$

$$CH_4 + H_2O \leftrightarrows CO + 3H_2 \qquad \Delta H^{\circ}_{298k} = +226 \text{ kJ/mol} \qquad (3)$$

$$CH_4 + CO_2 \leftrightarrows 2CO + 2H_2 \qquad \Delta H^{\circ}_{298k} = +261 \text{ kJ/mol} \qquad (4)$$

Moreover, some side reactions, such as water gas shift reaction (Equation (5)) and Boudouard reaction (Equation (6)) can also occur along with main reactions.

$$CO + H_2O \leftrightarrows CO_2 + H_2 \qquad \Delta H^{\circ}_{298k} = -41.2 \text{ kJ/mol} \qquad (5)$$

$$2CO \leftrightarrows C + CO_2 \qquad \Delta H^{\circ}_{298k} = -172.8 \text{ kJ/mol} \qquad (6)$$

The water gas shift and Boudouard reactions are exothermic in nature and take place at lower temperature. However, the respective reverse reactions occur upon increasing the reaction temperature.

Among the efficient catalysts for partial oxidation of methane (POM) are transition metals such as Ni, Pt, and Co supported on alumina, zirconia etc. However, these catalysts deactivate as a result of carbon formation [7,8]. It has been established that the activity of the Ni and/or Co catalysts not only relies on the structure and the nature of the active metals but selection of the support also plays a significant role. Al_2O_3 is extensively utilized as a support for reforming reactions. However, when Al_2O_3 is employed alone as a support for such type of catalysts, problems arise such as carbon deposition on active sites and development of inactive spinel phase ($NiAl_2O_4$) [9]. The modification of support, therefore, can be a promising route to enhance the catalytic performance. Among the prevalent materials, ZrO_2 has drawn considerable attention due to its excellent characteristics like acid-base properties, oxygen storage capacity and thermal stability [10]. It also inhibits the formation of spinels like $NiAl_2O_4$ by impeding the incorporation of active species into Al_2O_3 lattice [11,12]. Tetragonal zirconia is unstable at ambient temperature, but it can be stabilized by addition of Al_2O_3 to ZrO_2. Moreover, this binary system has a higher modulus of elasticity compared to neat ZrO_2 [11,13].

Several studies have been carried out on the formation of synthesis gas by using Ni and Co-based catalysts. Zagaynov et al. [14] examined Ni (Co)–Gd0.1Ti0.1Zr0.1Ce0.7O2 mesoporous catalysts obtained by co-precipitation for partial oxidation and dry reforming of methane. Surprisingly, the results showed that Co and Ni–Co-containing catalysts were more active in partial oxidation of methane than the Ni sample, while Ni-catalysts were more active in dry reforming of methane. Calcination temperature, on the other hand, affects the active metal particle size and therefore alters the stability of the catalysts by changing the diffusion path. Moreover, the calcination temperature has a significant impact on the structural and catalytic properties of the catalysts, which interact strongly with the metal oxide support. Other researchers [15,16] also highlighted the effect of pretreatment of catalysts at calcination temperature. On the other hand, other studies have demonstrated comparable performance at high temperatures or by using precious metals. For instance, Dedov and co-workers utilized neodymium-calcium cobaltate-based catalysts for syngas production via partial oxidation of methane [17]. They reported to attain 85% methane conversion and selectivity of CO and H_2 close to 100% at very a high temperature (925 °C). Likewise, another study used Ni(Co)-Gd0.1Ti0.1Zr0.1Ce0.7O2 catalyst and obtained comparable H_2 selectivity at a higher temperature (900 °C) for the production of syngas via partial oxidation of methane [14]. The present work is driven by our previous work [18] where it is has been shown that by using a single catalysis system of cobalt over CeO_2 and ZrO_2 supports; the hydrogen yield only up to 60% and 75 respectively was achieved for this system. Moreover, CeO_2 support yield low hydrogen and cobalt alone is considered less reforming catalysis. Therefore, in this work, the effect of binary metal system and support has been studied. It was observed that this system performs much better than single catalyst where hydrogen production was achieved up to 100%. Several studies have employed Co-based catalysts for reforming reactions [14,19,20]. For instance, Zagaynov et al. [14] examined (Ni, Co and Co-Ni)/-Gd0.1Ti0.1Zr0.1Ce0.7O2 mesoporous catalysts obtained by co-precipitation for partial oxidation and dry reforming of methane. Interestingly, the results showed that the Co- and Ni–Co- containing catalysts exhibited excellent catalytic performance in partial oxidation of methane than the Ni sample, while the Ni-catalysts demonstrated tremendous catalytic performance in dry reforming of methane.

Accordingly, the significance of this research contribution was to obtain a high catalytic performance at relatively low temperature using mono and bimetallic Co and Ni supported on (ZrO_2 + Al_2O_3) which are capable of producing syngas via partial oxidation of methane. In addition, they must be stable to overcome the deactivation processes like carbon accumulation, metal agglomeration and thermal sintering. The study of catalyst design started with a systematic investigation of the desired reaction together with potential side reactions. The sol-gel method of preparation was proposed to generate strong metal-support interaction (MSI) and to produce smaller metal particles, which is expected to be active in the catalytic reaction.

2. Materials and Methods

2.1. Materials

The chemicals used in the present study were all of analytical grade and supplied by Aldrich, Gillingham, UK. They included cobalt acetate Co(ac)$_2$·4H$_2$O, nickel acetate Ni(ac)$_2$·4H$_2$O and zirconium(IV)-butoxide Zr(BuO)$_4$ (80 wt % solution in 1-butanol). Aluminum tri-sec.-butylate Al(sec.-BuO)$_3$ was supplied by Merck, Southampton, UK.

2.2. Catalyst Preparation

The known sol-gel methods were adapted for the preparation of the catalysts. Precursors Co and Ni acetates were thoroughly dried to eliminate the moisture content. Then they were ground and sieved to obtain particle sizes <100 μm. The total metal loading is 5 wt % of Co and/or Ni in the monometallic catalyst, while for the bimetallic the total metal loading was 5 wt % with 1:1 mole ratio. The Zr to Al atoms is also 1:1 mole ratio. For the preparation of 16.33 g of ZrO_2–Al_2O_3 with an equimolar ratio of Zr to Al, 48 g of Zr-butylate (equivalent to 11.32 g of ZrO_2) and 25 g of Al-sec.-butylate (equivalent to 5.01 g of Al_2O_3) were placed in a 250 mL three-necked round glass bottom flask. The mixture was heated with continuous stirring to 130 °C. A lot of 2.59 g of dried Co acetate was added and the mixture was again heated for about two hours at the same temperature.

After completion, the reaction mixture was transferred into 75 g of isopropanol. A homogeneous solution was obtained with slightly pink color when using Co while it was faint green in presence of Ni. To this solution, 27 mL of distilled water were added immediately and the mixture was then refluxed for another hour. After cooling to room temperature, the precipitate was separated from the liquid with a glass frit. The obtained solid was first dried at room temperature overnight and then divided into two parts. One part was calcined under air at 550 °C for 5 h with a heating rate of 2 K/min. The other part of the solid was calcined at 800 °C under similar conditions. For simplicity, the catalyst names refer to their pre-treatment calcination temperature.

2.3. Catalyst Testing

Catalyst activity measurements were carried out using a Process Integral Development Engineering and Technology (PID Eng & Tech) Microactivity Setup equipped with a tubular stainless steel fixed-bed reactor (9 mm I.D., Autoclave Engineers, Pennsylvania, USA). The effluent gases were analyzed by an on-line gas chromatograph (GC, ALPHA MOS instrument, Toulouse, France) with a thermal conductivity detector at an interval of 30 min. For separation of the products, two GC columns Molecular Sieve 5A and Porapak Q were employed in series/bypass connections. A catalyst load of 0.15 g was used for each run while the total gas flow was fixed at 15 mL/min. Prior to the reaction the catalyst was reduced by dosing H_2 at a flow rate of 40 mL/min. The temperature was kept at 800 or 700 °C and held for 1 h in order to reduce the metal oxide into the active metal. Afterwards, the reactor was purged with N_2 till the required reaction temperature was achieved. The feed is not introduced to the reactor unless H_2 is completely removed from the system. This is done using GC analysis via TCD detector. A propak Q and molecular sieve columns were used for separation. The volume ratio of feed gases (CH_4/O_2) was set to 2. In addition, the space velocity was held at 6000 mL/(h·g_{cat}), while the total feed rate was set to 15 mL/min. The reaction temperature was checked by placing a thermocouple in the middle of the catalyst bed and the bed height was 0.4 cm. The reforming activity of catalysts was studied at 700 and 800 °C at 1 bar.

The composition of effluent gases was calculated by the normalization method, and the equations for determination of conversion and selectivity are used as following:

$$\text{Conversion of } CH_4 : X_{CH4} = \frac{CH_4 \text{ in} - CH_4 \text{ out}}{CH_4 \text{ in}} \times 100\% \qquad (7)$$

$$\text{Selectivity of } H_2 : S_{H2} = \frac{\text{moles of } H_2 \text{ produced}}{\text{Total moles of products } (H_2 + H_2O)} \times 100\% \qquad (8)$$

$$\text{Selectivity of } CO : S_{CO} = \frac{\text{moles of CO produced}}{\text{Total moles of products } (CO + CO_2)} \times 100\% \qquad (9)$$

2.4. Catalyst Characterization

Powder X-ray diffraction (XRD) analysis of fresh catalyst was conducted by employing a Rigaku (Miniflex) diffractometer with a Cu Kα1 radiation (λ = 0.15406 nm) operated at 40 mA and 40 kV. The 2θ range and scanning step for analysis were 10–80° and 0.02°, respectively.

The N_2 adsorption and desorption data at −196 °C was analyzed for determining the specific surface area (BET) of the fresh catalysts by using Micromeritics Tristar II 3020 surface area analyzer. In order to get rid of other adsorbed gases and moisture, all samples were degassed before analysis. For each analysis, a load of 0.2–0.3 g of catalyst was used. The pore size distribution of catalysts was calculated from the adsorption branch of N_2 isotherm by using the Barrett, Joyner & Halenda (BJH) method.

Temperature-programmed hydrogen reduction (H_2-TPR) and temperature-programmed carbon dioxide desorption (CO_2-TPD) measurements were performed on a chemisorption device (Micromeritics AutoChem II).

A known amount of catalyst was pre-treated with high purity argon (Ar) at 150 °C for about half an hour for TPR analysis. Then, the samples were heated in an automatic furnace to 1000 °C at a steady heating rate of 10 K/min under 40 mL/min of H_2/Ar mixture (volume ratio = 10/90) at atmospheric pressure. The H_2 signal was monitored by a thermal conduction detector (TCD).

For TPD experiments, first the adsorption of carbon dioxide onto the samples was carried out for half an hour at 50 °C under 10%CO_2/He gas at 30 mL/min. Then, the CO_2 desorption was done by increasing the temperature at a rate of 10 K/min to 800 °C.

The scanning electron microscopy (SEM) was employed in order to investigate the surface morphology of the catalysts. The SEM images of the spent catalyst samples were taken by using

JSM-7500F (JEOL Ltd., Tokyo, Japan) scanning electron microscope. The TEM study was carried out at 200 kV with an aberration-corrected JEM-ARM200F (JEOL, Corrector: CEOS). The microscope is fitted with a JED-2300 (JEOL) energy-dispersive X-ray-spectrometer for chemical analysis.

Temperature-programmed oxidation (TPO) experiments were conducted to determine the carbon accumulation on the spent catalyst after prolonged activity tests. The samples recovered from partial oxidation were dried at 150 °C for half an hour under helium at 30 mL/min and then cooled to ambient temperature. Afterwards, the temperature was raised with a ramp of 10 K/min to 800 °C under 10% O_2/He at 30 mL/min.

The quantitative analysis of coke deposition on the spent catalysts was carried out using thermo-gravimetric analyzer (Shimadzu, Kyoto, Japan). The spent catalysts weighing 10–15 mg were heated from ambient temperature to 1000 °C at a heating rate of 20 K/min, and the weight loss was recorded. For this purpose, catalyst samples recovered after 5 h on stream at 700 and 800 °C as well as Co-800 after long term test (24 h) at 800 °C were used. All analyses were carried out under air atmosphere.

3. Results and Discussion

3.1. X-ray Diffraction (XRD)

Typical XRD patterns in the range 2θ = 10–80° of fresh cobalt and/or nickel catalysts supported on the composite support (Al_2O_3 + ZrO_2) calcined at 550 and 800 °C are presented in Figure 1. In the case of samples calcined at 550 °C, broad reflections are observed. It is not possible to distinguish the species due to broadening and superimposing of reflections. Therefore, it implies that metal species are made of smaller crystallites and are well dispersed on the supports, which makes them amorphous and insensitive to X-ray radiations. This finding is consistent with the results obtained by BET and TPR which will be discussed later. Also, it is well known that the addition of zirconia to alumina leads to signal enlargement as a result of the formation of smaller particles [18,21]. Moreover, the decline in the intensity of the diffraction signals of catalysts Ni-550, Co-550 and Ni–Co-550 may also be caused by the distortion or defects in the Al-O bonds due to Zr presence in the support [22].

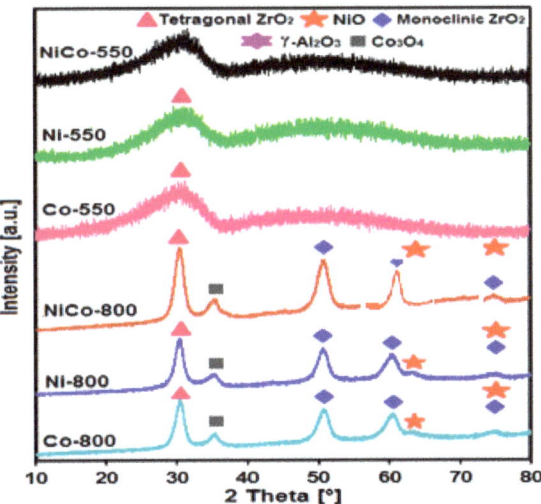

Figure 1. XRD patterns for fresh Ni and/or Co-based catalysts calcined at 550 and 800 °C.

With regard to the catalysts calcined at 800 °C, diffraction signals of sharp intensity observed; those represent more crystalline phases. Furthermore, the reflex intensity of the bimetallic catalyst is higher than for the monometallic catalysts. For the Ni-800 catalyst, the reflections obtained at 2θ = 63°, 75.3° and 79.4° are attributed to cubic NiO phase (JCPDS 01-73-1519). Actually, it is hard to identify the nickel oxide in the catalysts because its reflexes coincide with those of the tetragonal phase of zirconia [10]. The reflections observed at 2θ = 50.2°, 59.9°, 62.8° and 75.2° are ascribed to monoclinic ZrO_2 (JCPDS: 00-007-0343). The signals found at 2θ = 60.5° may be assigned to γ-Al_2O_3 (JCPDS: 00-029-0063). Only in case of Ni–Co-800, extra peaks detected at 2θ = 65.53° and 66.4° correspond to the formation of $NiAl_2O_4$ spinel phase. It is noteworthy that for both mono- and bimetallic catalysts the increase of the calcination temperature increases the reflex intensity which may be attributed to the formation of larger crystal size.

3.2. Textural Properties

The surface texture was assessed by using the nitrogen adsorption–desorption isotherms. Figure 2 illustrates the adsorption isotherms of the fresh catalysts calcined at 550 °C and 800 °C, while BET surface area, average pore diameter and pore volume are tabulated in Table 1. As per the IUPAC classification, catalysts demonstrate Type II isotherms. In Figure 2a it can be found that the BET surface area of Co-550 is highest and that of Ni-550 is lowest, while the surface area of Ni–Co-550 takes an intermediate value. On the other hand, the catalysts calcined at 800 °C (Figure 2b) exhibited a similar trend (Co > Ni–Co > Ni), however, the surface area of these catalysts was lower compared to those calcined at 550 °C, which may be due to the sintering. In a previous study, we showed that the addition of ZrO_2 to Al_2O_3 increased the surface area. For instance, the surface area of pure supported Ni/ZrO_2 and Ni/Al_2O_3 catalysts were 3.1 m^2/g and 122.0 m^2/g, respectively [23]. However, the surface area of binary supported Ni/Al_2O_3 + ZrO_2 catalyst had risen to 212 m^2/g.

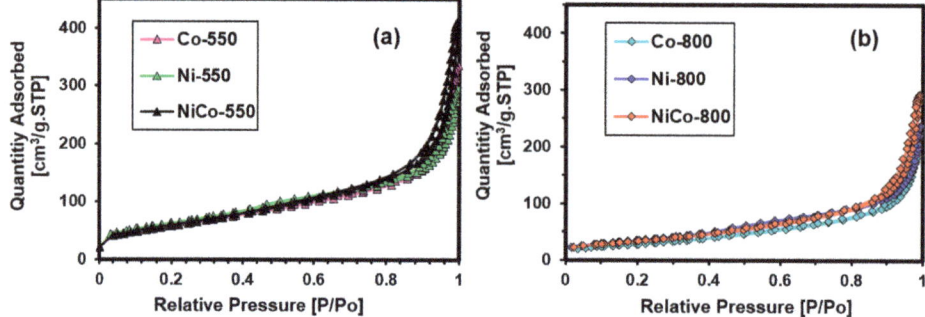

Figure 2. N_2 adsorption-desorption isotherms for fresh Ni and/or Co-based catalysts (**a**) calcined at 550 °C and (**b**) calcined at 800 °C.

Table 1. BET surface area, pore volume (P.V.) and pore diameter (P.D.) of fresh Ni and/or Co-based catalysts calcined at 550 °C and 800 °C.

Catalysts	BET Surface Area (m^2/g)	P.V. (cm^3/g)	P.D. (Å)
Ni-550	212	0.51	100
Co-550	230	0.44	82
Ni–Co-550	216	0.61	114
Ni-800	105	0.37	134
Co-800	130	0.38	104
Ni–Co-800	123	0.47	142

3.3. Temperature-Programmed Reduction (H_2-TPR)

In order to evaluate the reducibility of the species present in the catalyst, temperature-programmed reduction with hydrogen was employed. As shown in Figure 3a,b, Ni and/or Co catalysts calcined at

550 °C and 800 °C undergo a single-step reduction. NiO and CoO/Co$_3$O$_4$ are reducible species and can be categorized on the basis of their reduction temperature. As found in literature, bulk NiO is reduced between 300 and 400 °C [24,25]. In the case of the catalysts calcined at 550 °C (Figure 3a), a broad and pronounced reduction peak is observed for Ni-550 at 450–700 °C with a peak centered at 590 °C. This indicates that the Ni^{2+} species are difficult to reduce due to their interaction with the support (forming spinel) [26].

It is well known that the TPR profiles of cobalt catalysts demonstrate two distinct metal oxide species being reduced at specific temperature. First, region (<400 °C) is assigned to the reduction of Co$_3$O$_4$ to CoO. The second region (400–500 °C) corresponds to the reduction of CoO to metallic Co^0 [27]. Therefore, the H$_2$ reduction peak with a maximum around 870 °C can be attributed to the reduction of Co^{2+} species with strong support interaction.

With regard to the bimetallic Ni–Co-550 catalyst, the combination of both Ni and Co enhances the reducibility of Co. The broadening of the reduction peak in the high-temperature zone may be attributed to the reduction of Ni–Co$_2$O$_4$ species forming Co–Ni alloy, having strong interaction with Al$_2$O$_3$ + ZrO$_2$ as proposed by our previous study [28]. It is noteworthy that the TPR peak area of Ni–Co-550 is higher than the monometallic one, suggesting that it possesses more reducible species. On the other hand, H$_2$-TPR conducted for the catalysts calcined at 800 °C (Figure 3b) followed the same trend. However, the position and the intensity of the peaks were different. In the case of Co-800, the single reduction peak may be designated to the overlapping of two-stage reduction of Co$_3$O$_4$ → CoO → Co metal [29,30]. The shift of reduction peaks suggests the existence of strong interaction between Co^{2+} and support due to calcination. The same shift was also observed for both Ni and bimetallic Ni–Co. Moreover, Co-800 was found to have the highest intensity and peak shift as it was found for the samples calcined at 550 °C.

In our system ZrO$_2$ obviously doesn't interact with Al$_2$O$_3$ strongly and the interaction between ZrO$_2$ and Ni is weak as was found by J. Asencios et al. [31] so Ni and/or Co is able to interact with Al$_2$O$_3$ to form Ni and/or CoAl$_2$O$_4$ and by this the Ni and/or Co are highly dispersed and as they are small crystals it is not possible to observe by XRD. The extent of this transformation increased with calcination temperature and is evidenced by the shift in reduction temperature for samples calcined at different temperatures. Also, G. P. Berrocal et al. [10] found that Ni strongly interacts with aluminum forming small NiAl$_2$O$_4$ particles that have the highest reduction temperature. At the same time, this sample showed the highest catalytic activity for the partial oxidation of methane. We observed similar dependency in our results.

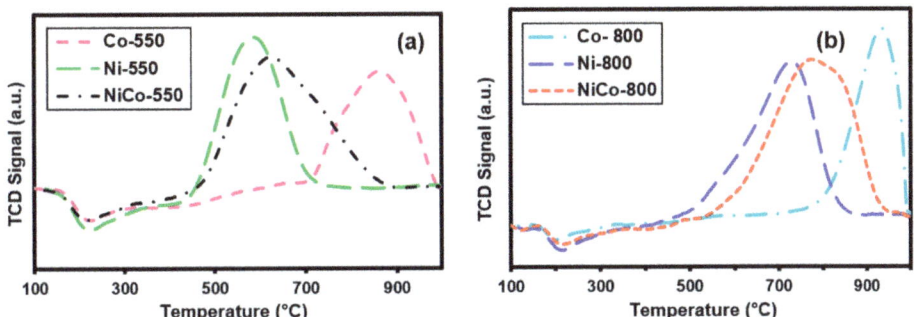

Figure 3. TPR profiles for fresh monometallic Ni or Co and bimetallic Ni–Co-based catalysts (**a**) calcined at 550 °C and (**b**) calcined at 800 °C.

3.4. Thermal Analysis for Carbon Deposition

TGA analysis was conducted to quantify the deposited carbon over the spent catalysts. In Figure 4a,b, the TGA profiles illustrate the weight loss (%) as a function of temperature for all recovered catalysts from tests at 700 and 800 °C, respectively. In general, the amount of deposited carbon was relatively low for all the tested catalysts due to the presence of zirconia which is well known for high oxygen storage capacity and the presence of basic centers. The relative carbon deposition after reaction at 700 °C can be assigned in the following order: Ni-800 ≈ Co-550 < Co-800 < Ni-550 < Ni–Co-800 < Ni–Co-550 (Figure 4b). For all catalysts, the burning of carbon starts at the same temperatures around 500 °C except for Ni–Co-550. From Figure 4 it is clear that Co-550 was the least prone to carbon deposition at both reaction temperatures 700 °C and 800 °C because cobalt is recognized as a strong oxidizing catalyst which can tackle the soot formation [32]. Interestingly, all catalysts calcined at 800 °C were found to have lower and similar amount of carbon deposits after reaction at 800 °C (encircled in Figure 4b), which can be associated to the strong interaction of metal species with composite support as it has been discussed in Section 3.3 [14]. Consequently, it can be deduced that higher calcination and reaction temperatures pose no adverse effect to our catalysts because they were less susceptible to carbon deposition. We assume that increasing the calcination temperature from 550 to 800 °C may form new surface sites due to the strong metal-support interaction. This might stabilize the high Ni and/or Co dispersion against metal agglomeration and deactivation. Apart from this, ZrO_2 might activate the oxidation of coke at high temperature and prevent the catalysts from coking. Also, Co-800 catalyst operated at 800 °C had excellent stability for 24 h on stream without deactivation (as it will be discussed latter). Henceforth, monometallic catalysts presented better performance with higher calcination temperature than bimetallic ones. In addition, the rate of coking over Ni–Co-800 was higher in comparison with mono-metallic catalysts which is consistent with the findings reported in the literature [33,34]. Moreover, this effect was more pronounced for the catalysts calcined at 550 °C (Figure 4a).

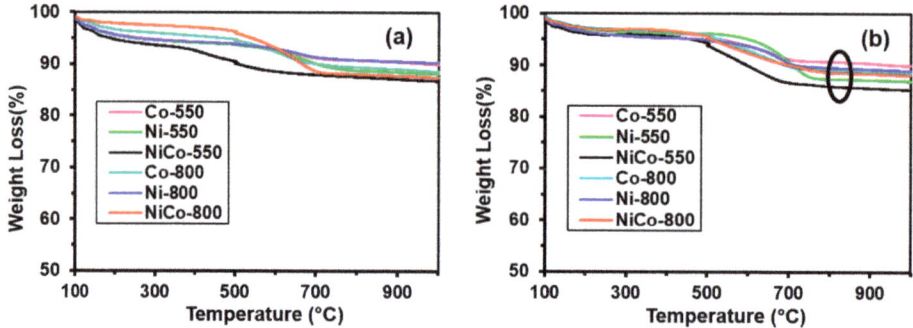

Figure 4. TGA profiles for spent Ni and/or Co-based catalysts calcined at 550 °C and at 800 °C after tests at (**a**) 700 °C and (**b**) 800 °C.

3.5. Temperature-Programmed Desorption of CO_2 (CO_2-TPD)

The basicity of the Ni and/or Co-containing catalysts was evaluated by adsorption and desorption of CO_2 on the basic sites at different temperatures. Figure 5 represents the CO_2-TPD profiles of the catalysts. The strength of basic sites can be classified by the temperature of the corresponding desorption peak of CO_2: weakly basic in the range of 50–200 °C, intermediate basic (200–400 °C), strongly basic (400–650 °C) and very strong basic sites (>650 °C) [35]. In fact, all these basic sites are evident from CO_2-TPD profiles, which reveal the strong basic character of the catalysts (Figure 5). Al_2O_3 as an acidic support favors coke formation. Therefore, ZrO_2 addition has rendered the catalysts basic character, which in turn escalated CO_2 adsorption contributing to higher activity and coke removal.

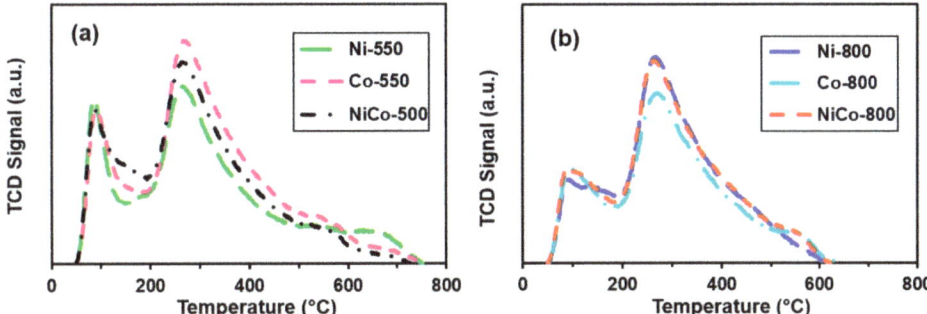

Figure 5. CO_2-TPD profiles for fresh Ni and/or Co-based catalysts after calcination at (**a**) 550 °C and (**b**) 800 °C.

3.6. Scanning Electron Microscopy (SEM) and Transmission Electron Microscopy (TEM)

Figure 6 displays SEM images of fresh and spent catalysts obtained after five hours on stream at 700 °C and corresponding samples calcined at 550 °C. The fresh catalyst surface shows a fairly good distribution of the particles while the spent catalyst shows agglomeration of the particles and therefore the surface area and Ni dispersion decrease. The catalytic activity is strongly affected by carbon deposition over the catalysts' surfaces, finally deactivating the catalyst.

TEM of Co/(Al_2O_3–ZrO_2) catalyst calcined at 800 °C used in the long term POM test at 800 °C reveals presence of filamentous coke and the size of carbon nanotubes (CNTs) is determined by the size of starting metallic species. These CNTs gradually grow and metallic Co species settled on the tip of the CNTs. As the metallic species are still exposed to the reacting gases, these CNTs do not show an adverse effect on activity because metallic species are still accessible [28].

Figure 6. SEM images of Co–Ni/(Al_2O_3–ZrO_2) catalyst. (**A**) fresh catalyst, (**B**) spent catalyst calcined at 800 °C and operated at 700 °C reaction for 5 h and (**C**) TEM of Co/(Al_2O_3–ZrO_2) catalyst calcined at 800 °C after recovery from long term POM test (24 h) at 800 °C.

3.7. Catalytic Activity

The product H2/CO ratio for all catalysts is slightly higher than the stoichiometric value of 2 (Figure 7), owing to the incomplete conversion of CO2 (from combustion, Equation (1)) to CO. Also, a part of CO was consumed by the side reactions such as water-gas shift (Equation (5)) and Boudouard reactions (Equation (6)). Consequently, both of these effects lower the CO selectivity (Figure 8b) with time on stream and thus increase the H2/CO ratio at 700 °C. Moreover, Co-800 gave the lowest CO2 selectivity (15.2%) and the highest CO selectivity (85%), eventually attained H2/CO ratio approaching the stoichiometric value of 2. It is worth to mention that the selectivity for hydrogen reached 98.6% for all catalysts at 700 °C.

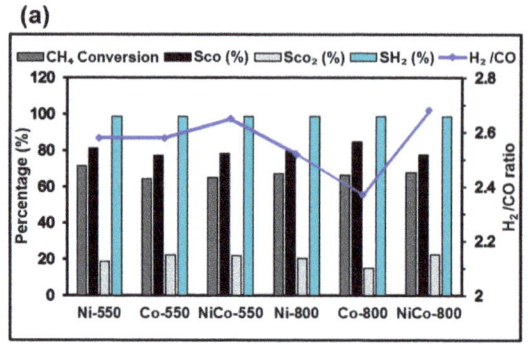

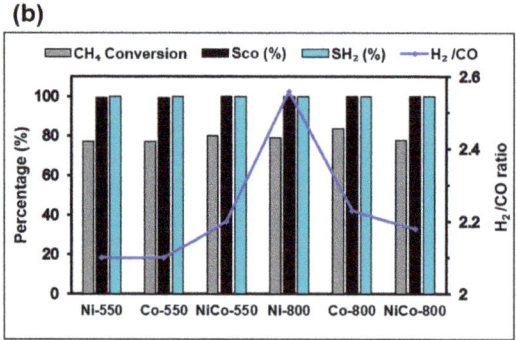

Figure 7. Conversion, selectivity and H$_2$/CO ratio obtained for Ni and/or Co-based catalysts operated at (**a**) 700 °C and (**b**) 800 °C.

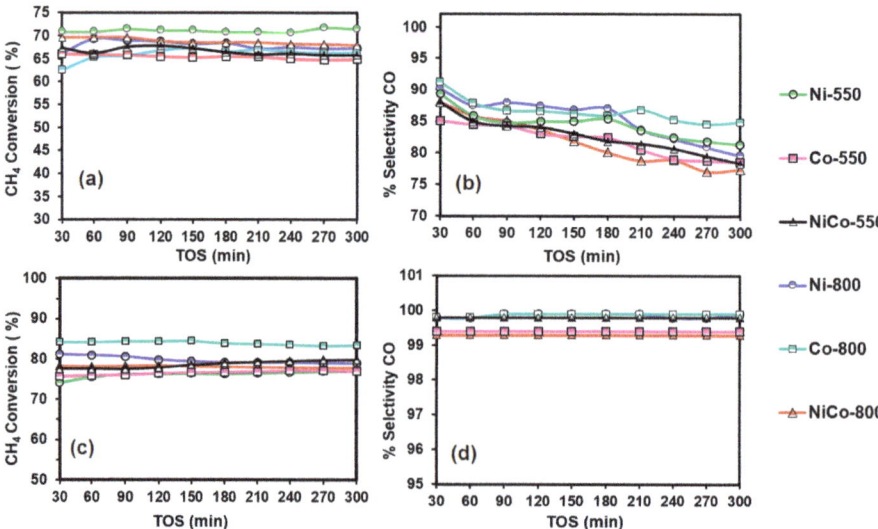

Figure 8. (**a**) CH$_4$ conversion and (**b**) CO selectivity with time on stream in POM over Ni and/or Co-based catalysts at 700 °C; (**c**) CH$_4$ conversion and (**d**) CO selectivity with time on stream in POM over Ni and/or Co-based catalysts at 800 °C.

The performance of the Ni/ZrO$_2$–Al$_2$O$_3$, Co/ZrO$_2$–Al$_2$O$_3$ and Co–Ni/ZrO$_2$–Al$_2$O$_3$ catalysts calcined at 550 °C and 800 °C was tested at 700 °C and 800 °C (Figure 8). Generally, the activity of catalysts progressively increases with rise in the reaction temperature. The oxygen conversion was unaltered (nearly 98%) for all catalysts irrespective of calcination temperatures. At 700 °C (Figure 8a) maximum conversion of 71.5% was achieved with the Ni-550 catalyst. This might be attributed to its high surface area compared to the other catalysts (Figure 2). It may also be due to minimum carbon deposition formed on this catalyst as shown in Figure 4. On the other hand, the activity of the bimetallic catalyst suffered from the formation of carbon deposits (Figure 4a,b). The lower activity of Ni–Co-800 catalyst may also be due to the formation of spinel phase as it was discussed with the help of TPR (Figure 1). These species are irreducible and do not contribute to methane conversion [36]. Since Al^{3+} and Ni^{2+} are located in the same lattice, the generation of the solid solution of NiAl$_2$O$_4$ spinel is conducive under higher calcination temperatures what about ZrO$_2$ [11]. Moreover, the highest selectivity for H$_2$ of 99% is achieved with all the catalysts when operated at 700 °C. As steam reforming (Equation (3)) is thermodynamically feasible at this temperature and water is available, it contributes to the rise in selectivity to H$_2$.

Generally, when the partial oxidation was carried out at 800 °C, both CH$_4$ conversion and CO selectivity were remarkably increased (Figure 7b). Moreover, methane conversion for all catalysts was found to be in the order: Co-800 > Ni–Co-550 > Ni-800 > Ni–Co-800 ≈ Ni-550 = Co-550. The selectivities to CO and H$_2$ achieved with all the catalysts exceeded 99% at 800 °C. In addition, the amount of CO$_2$ was minimum (<1%) in the product stream which implies that CO$_2$ has been converted into CO. All the catalysts maintained their activity throughout the test duration, which can be associated to the higher calcination temperature and the presence of ZrO$_2$. An intimate contact with metal species is developed by the presence of ZrO$_2$ due to strong electrostatic attraction between them. This fact is also evident from the TPR profiles. Furthermore, the strong metal-support interaction in these catalysts is responsible for the low carbon deposition.

Interestingly, when comparing to other catalysts, it was found that the monometallic Co-800 was the most active (84% CH$_4$ conversion) and stable catalyst (Figure 8c). In the presence of ZrO$_2$, the interaction between Al$_2$O$_3$ and Ni and/or Co increases. Ni and/or Co deposit on the support and develop an intimate contact which results in the modification of Al$_2$O$_3$ support [6]. The TPR of the monometallic sample Co-800 showed the highest reduction temperature due to the formation of stable spinel structures with the support. These interactions probably assist in dispersing the metals and coke formation resistance. The slight decline in the methane conversion of Ni-800 may be ascribed to blocking of active sites by carbon deposits (Figure 4) and relatively lower basicity (Figure 5). Similarly, at 800 °C the selectivity to CO remained constant throughout the stability test for all tested catalysts (Figure 8d). Consequently, the rise in the CO selectivity at 800 °C shifted the H$_2$/CO ratio to a value closer to 2. It is worth to mention that the reaction temperature of 800 °C is most favorable for reduction of the tested metal oxides as can be seen in TPR profiles (Figure 3). A similar study was conducted using the same catalyst Ni/(ZrO$_2$ + Al$_2$O$_3$) but employing a higher metal loading (8%) and a calcination temperature of 550 °C. The catalyst achieved almost comparable methane conversions, but higher amount of carbon deposits and significantly lower selectivity to CO and H$_2$ [10]. The comparison of this result with the present study suggests that calcination temperature has a significant influence on the catalytic performance.

The higher activity of bimetallic catalyst Ni–Co-550 can be attributed to the synergistic effect between Ni and Co which is in agreement with several findings [14]. This effect induces higher BET surface area, smaller crystallite size (XRD) and improved degree of reducibility (TPR). Co- and Ni–Co-based catalysts presented higher catalytic activity than Ni-based catalysts. This finding is consistent with recent studies conducted by Zagaynov and co-workers using (Ni, Co and Co–Ni)/–Gd$_{0.1}$Ti$_{0.1}$Zr$_{0.1}$Ce$_{0.7}$O$_2$ catalysts [14]. However, the decline in activity of Ni–Co-800 calcined at 800 °C may be ascribed to the formation of spinel phases as described above. On the basis of catalytic activity, Co-800 is the most promising catalyst giving higher conversion and excellent selectivity to CO (85%) as well as H_2 (98.6%) even at 700 °C, and this selectivity can reach 100% at 800 °C reaction temperature. Therefore, it is evident that the monometallic catalysts gave better performance with higher calcination temperature while bimetallic catalysts exhibit higher activity with lower calcination temperature.

3.8. Long-Term Stability Test

Generally, catalyst stability in POM is greatly influenced by deactivation resulting from sintering, metal agglomeration, carbon deposition, and the disappearance of active sites due to oxidation at reaction conditions. Usually, these deactivation effects occur simultaneously, but sometimes one of them predominates. Among the catalysts used in this study, Co-800 showed best results and so it was selected for a prolonged activity test at 800 °C for 24 h (Figure 9). It is worth to mention that the catalyst maintained stable activity throughout the complete run.

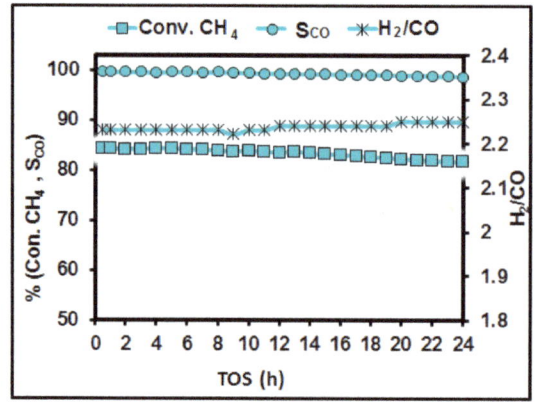

Figure 9. CH_4 conversion, CO selectivity and H_2/CO ratio over Co-800 (5%Co/Al_2O_3–ZrO_2) catalyst calcined at 800 °C over 24 h on stream in POM at 800 °C.

The stable activity may be attributed to the presence of ZrO_2 that leads to coke suppression as revealed by TGA and TPO (Figure 10). The presence of ZrO_2 imparts two advantages to the catalysts: (i) It renders basic character (Figure 5) to the catalysts which in turn makes it capable of activating CO_2 ($CO_2 \rightarrow CO + O^*$) because it enhances the dissociative chemisorption of CO_2 in metal/ZrO_2 interface; (ii) it suppresses the carbon deposition as an outcome of its higher oxygen storage capacity which provides more active oxygen species by redox activity ($C^* + O^* \rightarrow CO$). This is the reason why the catalysts showed very low coking, making them long-term stable. Similar studies were conducted using Pt/Al_2O_3–ZrO_2 and Ni/Al_2O_3–ZrO_2 catalysts; higher activity and stability to syngas were reported [37,38]. This behavior is due to the rise in capacity of dissociative chemisorption of CO_2 over Pt-ZrO_2 and Ni-ZrO_2. Therefore, based on the stability analysis, it can be concluded that the catalyst operated at 800 °C was more stable than the one tested at 700 °C (Figure 8a–d).

3.9. Post (Long Term Test) Characterizations

Temperature-programmed oxidation (TPO) was conducted to characterize the nature of coke deposit over Co-800 catalyst after the long-term POM test (Figure 10a). Zhang investigated the TPO profiles for reforming reaction and assigned three peaks as $C\alpha$ (150–220 °C), $C\beta$ (530–600 °C) and $C\gamma$ (~650 °C) whereas the peak above 700 °C might indicate the oxidation of graphitic/inactive carbon [39]. We applied this model to our catalysts. As per TPO profile, the intensity maxima of $C\alpha$ was found at 293 °C corresponding to the most active carbon which is responsible for the transformation into synthesis gas. The maximum at 593 °C represents $C\beta$ which may be attributed to intermediate amorphous carbon and could be transformed into CO at high temperature. Finally, the peak at 665 °C possessing the lowest intensity may be ascribed to Cγ, an inert carbon intermediate which is transformed into filamentous or graphitic features. The intensity of the signal for the most active carbon (Cα) is higher which implies that these species are predominant. These findings are in agreement with TEM images (Figure 6c). When TGA (Figure 10b) was performed after the test over 24 h with Co-800 catalyst at 800 °C, it was found that there was insignificant (<1%) rise in the coke amount on the catalyst surface. The low amount of carbon may be attributed to the much amount of active and amorphous carbon type which is also registered by TPO.

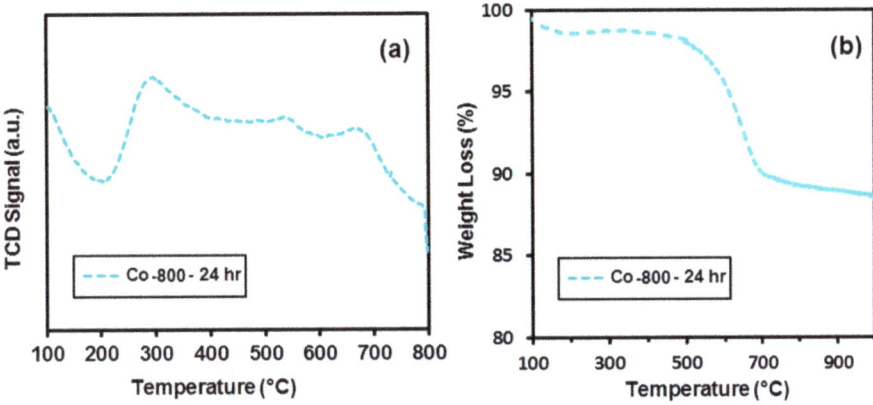

Figure 10. (a) Temperature-programmed oxidation and (b) TGA for Co-800 (5%Co/Al$_2$O$_3$—rO$_2$) catalyst calcined at 800 °C after the long-term POM test at 800 °C.

4. Conclusions

The obtained results show that the ZrO$_2$–Al$_2$O$_3$-supported Ni and/or Co catalysts for syngas production via partial oxidation exhibit a high surface area. Co/Al$_2$O$_3$–ZrO$_2$ catalysts demonstrated superior catalytic performance, giving high methane conversion and selectivity to CO and H$_2$ at 700 °C and reached up to 100% selectivity to H$_2$ and 84% methane conversion at 800 °C. Increasing the calcination temperature from 550 °C to 800 °C resulted in strong metal-support interaction which endowed resistance against sintering. The presence of ZrO$_2$ in the binary oxide enhanced the surface area and number of basic sites in the catalysts. Several factors can assist to obtain stable and active catalysts as the presence of basic sites by addition of ZrO$_2$-facilitated CO$_2$ dissociation, generation of oxygen intermediates, and removal of deposited carbon over the catalyst surface. Furthermore, the effect of calcination at a higher temperature of 800 °C stabilizes high dispersion of Ni and/or Co on the support, thereby avoiding metal agglomeration which in turn improved coke resistance. Eventually, monometallic Co-based catalyst calcined at 800 °C was found to have the highest activity but not Ni-based catalyst, which is unexpected. On the other hand, bimetallic Ni–Co-550 showed highest activity at low calcination temperature. Finally, increasing the calcination and reaction temperatures led to higher activity but posed no adverse effects on stability. It is worth mentioning that Co-800 catalyst used at 800 °C was found to have excellent stability over 24 h on stream. Recently, Dedov and co-workers utilized neodymium-calcium cobaltate-based catalysts for syngas production via partial oxidation of methane by using a fixed-bed flow reactor [17]. They reportedly attained 85% methane conversion and selectivity of CO and H$_2$ close to 100% at very high temperature (925 °C). Another study used Ni(Co)–Gd$_{0.1}$Ti$_{0.1}$Zr$_{0.1}$Ce$_{0.7}$O$_2$ catalyst at 900 °C for the production of syngas via partial oxidation of methane [14]. They obtained 80–90% methane conversion, 85–95% selectivity for CO and 79% selectivity for H$_2$. Methane conversion was somewhat higher but the selectivity to CO and H$_2$

was still lower even at a higher temperature. Based on the activity of catalysts reported in previous studies, our catalysts showed higher activity and selectivity at lower temperature.

Author Contributions: A.S.A.-F., A.H.F. and Y.A. carried out all experiments and characterization tests as well as shared in the analysis of the data and writing of the manuscript. U.A., H.A. and A.E.A. wrote the paper and shared data analysis. H.S. and A.A.I. contributed in writing the paper and edited it.

Funding: The research is funded by Deanship of Scientific Research at King Saud University project No. RGP-119.

Acknowledgments: The authors would like to extend their sincere appreciation to the Deanship of Scientific Research at King Saud University for funding this research group No. RGP-119.

Conflicts of Interest: The authors declare no conflict of interest.

References

1. Liu, Y.; Wang, T.; Li, Q.; Wang, D. A Study of Acetylene Production by Methane Flaming in a Partial Oxidation Reactor. *Chin. J. Chem. Eng.* **2011**, *19*, 424–433. [CrossRef]
2. Chibane, L.; Djellouli, B. Role of Periodic Input Composition and Sweeping Gas for Improvement of Hydrogen Production in a Palladium Membrane Reactor by Partial Oxidation of Methane. *Chin. J. Chem. Eng.* **2012**, *20*, 577–588. [CrossRef]
3. Dong, X.; Zhang, H.; Lin, W. Preparation and Characterization of a Perovskite-type Mixed Conducting $SrFe_{0.6}Cu_{0.3}Ti_{0.1}O_{3-\delta}$ Membrane for Partial Oxidation of Methane to Syngas. *Chin. J. Chem. Eng.* **2008**, *16*, 411–415. [CrossRef]
4. Fakeeha, A.H.; Al-Fatesh, A.S.; Chowdhury, B.; Ibrahim, A.A.; Khan, W.U.; Hassan, S.; Sasudeen, K.; Abasaeed, A.E. Bi-metallic catalysts of mesoporous Al_2O_3 supported on Fe, Ni and Mn for methane decomposition: Effect of activation temperature. *Chin. J. Chem. Eng.* **2018**, *26*, 1904–1911. [CrossRef]
5. Jun, J.H.; Lee, S.J.; Lee, S.H.; Lee, T.J.; Kong, S.J.; Lim, T.H.; Nam, S.W.; Hong, S.A.; Yoon, K.J. Characterization of a nickel-strontium phosphate catalyst for partial oxidation of methane. *Korean J. Chem. Eng.* **2003**, *20*, 829–834. [CrossRef]
6. Nichio, N.; Casella, M.; Ferretti, O.; Gonzalez, M.; Nicot, C.; Moraweck, B.; Frety, R. Partial oxidation of methane to synthesis gas. Behaviour of different Ni supported catalysts. *Catal. Lett.* **1996**, *42*, 65–72. [CrossRef]
7. Singha, R.K.; Shukla, A.; Yadav, A.; Konathala, L.S.; Bal, R. Effect of metal-support interaction on activity and stability of $Ni-CeO_2$ catalyst for partial oxidation of methane. *Appl. Catal. Environ.* **2017**, *202*, 473–488. [CrossRef]
8. Ding, C.; Ai, G.; Zhang, K.; Yuan, Q.; Han, Y.; Ma, X.; Wang, J.; Liu, S. Coking resistant $Ni/ZrO_2@SiO_2$ catalyst for the partial oxidation of methane to synthesis gas. *Int. J. Hydrog. Energy* **2015**, *40*, 6835–6843. [CrossRef]
9. Wu, P.; Li, X.; Ji, S.; Lang, B.; Habimana, F.; Li, C. Steam reforming of methane to hydrogen over Ni-based metal monolith catalysts. *Catal. Today* **2009**, *146*, 82–86. [CrossRef]
10. Berrocal, G.P.; Da Silva, A.L.; Assaf, J.M.; Albornoz, A.; do Carmo Rangel, M. Novel supports for nickel-based catalysts for the partial oxidation of methane. *Catal. Today* **2010**, *149*, 240–247. [CrossRef]
11. Sharifi, M.; Haghighi, M.; Rahmani, F.; Karimipour, S. Syngas production via dry reforming of CH_4 over Co-and Cu-Promoted $Ni/Al_2O_3-ZrO_2$ nanocatalysts synthesized via sequential impregnation and sol-gel methods. *J. Nat. Gas Sci. Eng.* **2014**, *21*, 993–1004. [CrossRef]
12. Therdthianwong, S.; Siangchin, C.; Therdthianwong, A. Improvement of coke resistance of Ni/Al_2O_3 catalyst in CH_4/CO_2 reforming by ZrO_2 addition. *Fuel Process. Technol.* **2008**, *89*, 160–168. [CrossRef]
13. Song, J.H.; Han, S.J.; Yoo, J.; Park, S.; Kim, D.H.; Song, I.K. Hydrogen production by steam reforming of ethanol over $Ni-X/Al_2O_3-ZrO_2$ (X = Mg, Ca, Sr, and Ba) xerogel catalysts: Effect of alkaline earth metal addition. *J. Mol. Catal. Chem.* **2016**, *415*, 151–159. [CrossRef]
14. Zagaynov, I.; Loktev, A.; Arashanova, A.; Ivanov, V.; Dedov, A.; Moiseev, I. Ni (Co)-$Gd_{0.1}Ti_{0.1}Zr_{0.1}Ce_{0.7}O_2$ mesoporous materials in partial oxidation and dry reforming of methane into synthesis gas. *Chem. Eng. J.* **2016**, *290*, 193–200. [CrossRef]

15. Weng, W.Z.; Pei, X.Q.; Li, J.M.; Luo, C.R.; Liu, Y.; Lin, H.Q.; Huang, C.J.; Wan, H.L. Effects of calcination temperatures on the catalytic performance of Rh/Al_2O_3 for methane partial oxidation to synthesis gas. *Catal. Today* **2006**, *117*, 53–61. [CrossRef]
16. Sokolov, S.; Kondratenko, E.V.; Pohl, M.-M.; Rodemerck, U. Effect of calcination conditions on time on-stream performance of Ni/La_2O_3-ZrO_2 in low-temperature dry reforming of methane. *Int. J. Hydrog. Energy* **2013**, *38*, 16121–16132. [CrossRef]
17. Dedov, A.G.; Loktev, A.S.; Komissarenko, D.A.; Parkhomenko, K.V.; Roger, A.C.; Shlyakhtin, O.A.; Mazo, G.N.; Moiseev, I.I. High-selectivity partial oxidation of methane into synthesis gas: The role of the redox transformations of rare earth-alkali earth cobaltate-based catalyst components. *Fuel Process. Technol.* **2016**, *148*, 128–137. [CrossRef]
18. Abasaeed, A.E.; Al-Fatesh, A.S.; Naeem, M.A.; Ibrahim, A.A.; Fakeeha, A.H. Catalytic performance of CeO_2 and ZrO_2 supported Co catalysts for hydrogen production via dry reforming of methane. *Int. J. Hydrog. Energy* **2015**, *40*, 6818–6826. [CrossRef]
19. Enger, B.C.; Lødeng, R.; Anders Holmen, A. Modified cobalt catalysts in the partial oxidation of methane at moderate temperatures. *J. Catal.* **2009**, *262*, 188–198. [CrossRef]
20. Lødeng, R.; Bjørgum, E.; Christian, B.; Enger Eilertsen, J.L.; Anders Holmen, A.; Krogh, B.; Rønnekleiv, M.; Erling Rytter, E. Catalytic partial oxidation of CH_4 to H_2 over cobalt catalysts at moderate temperatures. *Appl. Catal.* **2007**, *333*, 11–23. [CrossRef]
21. Silver, R.G.; Hou, C.J.; Ekerdt, J.G. The role of lattice anion vacancies in the activation of CO and as the catalytic site for methanol synthesis over zirconium dioxide and yttria-doped zirconium dioxide. *J. Catal.* **1989**, *118*, 400–416. [CrossRef]
22. Li, G.; Li, W.; Zhang, M.; Tao, K. Morphology and hydrodesulfurization activity of CoMo sulfide supported on amorphous ZrO_2 nanoparticles combined with Al_2O_3. *Appl. Catal.* **2004**, *273*, 233–238. [CrossRef]
23. Naeem, M.A.; Al-Fatesh, A.S.; Abasaeed, A.E.; Fakeeha, A.H. Activities of Ni-based nano catalysts for CO_2-CH_4 reforming prepared by polyol process. *Fuel Process. Technol.* **2014**, *122*, 141–152. [CrossRef]
24. Klein, J.C.; Hercules, D.M. Surface characterization of model Urushibara catalysts. *J. Catal.* **1983**, *82*, 424–441. [CrossRef]
25. Mile, B.; Stirling, D.; Zammitt, M.A.; Lowell, A.; Webb, M. The location of nickel oxide and nickel in silica-supported catalysts: Two forms of "NiO" and the assignment of temperature-programmed reduction profiles. *J. Catal.* **1998**, *114*, 217–229. [CrossRef]
26. Kim, P.; Kim, Y.; Kim, H.; Song, I.K.; Yi, J. Synthesis and characterization of mesoporous alumina with nickel incorporated for use in the partial oxidation of methane into synthesis gas. *Appl. Catal.* **2004**, *272*, 157–166. [CrossRef]
27. Arone, S.; Bagnasco, G.; Busca, G.; Lisi, L.; Russo, G.; Turco, M. Catalytic combustion of methane over transition metal oxides. *Stud. Surf. Sci. Catal.* **1998**, *119*, 65–70.
28. Al-Fatesh, A.S.; Arafat, Y.; Ibrahim, A.A.; Atia, H.; Fakeeha, A.H.; Armbruster, U.; Abasaeed, A.E.; Frusteri, F. Evaluation of Co-Ni/Sc-SBA-15 as a novel coke resistant catalyst for syngas production via CO_2 reforming of methane. *Appl. Catal.* **2018**, *567*, 102–111. [CrossRef]
29. Jongsomjit, B.; Panpranot, J.; Goodwin, J.G. Effect of zirconia-modified alumina on the properties of Co/γ-Al_2O_3 catalysts. *J. Catal.* **2003**, *215*, 66–77. [CrossRef]
30. Jongsomjit, B.; Sakdamnuson, C.; Goodwin, J.G.; Praserthdam, P. Co-support compound formation in titania-supported cobalt catalyst. *Catal. Lett.* **2004**, *94*, 209–215. [CrossRef]
31. Asencios, Y.J.O.; Nascente, P.A.P.; Assaf, E.M. Partial oxidation of methane on NiO-MgO-ZrO_2 catalysts. *Fuel* **2012**, *97*, 630–637. [CrossRef]
32. Harrison, P.G.; Ball, I.K.; Daniell, W.; Lukinskas, P.; Céspedes, M.A.; Miró, E.E.; Ulla, M.A.A. Cobalt catalysts for the oxidation of diesel soot particulate. *Chem. Eng. J.* **2003**, *95*, 47–55. [CrossRef]
33. Rui, Z.; Feng, D.; Chen, H.; Ji, H. Anodic TiO_2 nanotube array supported nickel-noble metal bimetallic catalysts for activation of CH_4 and CO_2 to syngas. *Int. J. Hydrog. Energy* **2014**, *39*, 16252–16261. [CrossRef]
34. Djinović, P.; Črnivec, I.G.O.; Erjavec, B.; Pintar, A. Influence of active metal loading and oxygen mobility on coke-free dry reforming of Ni–Co bimetallic catalysts. *Appl. Catal.* **2012**, *125*, 259–270. [CrossRef]
35. Al-Fatesh, A.S.; Arafat, Y.; Atia, H.; Ibrahim, A.A.; Manh Ha, Q.L.; Schneider, M.; Pohl, M.M.; Fakeeha, A.H. CO_2-reforming of methane to produce syngas over Co-Ni/SBA-15 catalyst: Effect of support modifiers (Mg, La and Sc) on catalytic stability. *J. CO_2 Util.* **2017**, *21*, 395–404. [CrossRef]

36. Lu, Y.; Liu, Y.; Shen, S. Design of stable Ni catalysts for partial oxidation of methane to synthesis gas. *J. Catal.* **1998**, *177*, 386–388. [CrossRef]
37. Pompeo, F.; Nichio, N.N.; Ferretti, O.A.; Resasco, D. Study of Ni catalysts on different supports to obtain synthesis gas. *Int. J. Hydrog Energy* **2005**, *30*, 1399–1405. [CrossRef]
38. Souza, M.; Aranda, D.; Scmal, M. Reforming of Methane with Carbon Dioxide over $Pt/ZrO_2/Al_2O_3$ Catalysts. *J. Catal.* **2001**, *204*, 498–511. [CrossRef]
39. Zhang, Z.; Verykios, X. Carbon dioxide reforming of methane to synthesis gas over supported Ni catalysts. *Catal. Today* **1994**, *21*, 589–595. [CrossRef]

© 2019 by the authors. Licensee MDPI, Basel, Switzerland. This article is an open access article distributed under the terms and conditions of the Creative Commons Attribution (CC BY) license (http://creativecommons.org/licenses/by/4.0/).

Article

The Synthesis of N-(Pyridin-2-yl)-Benzamides from Aminopyridine and Trans-Beta-Nitrostyrene by Fe$_2$Ni-BDC Bimetallic Metal–Organic Frameworks

Trinh Duy Nguyen [1,2,*], Oanh Kim Thi Nguyen [1,2], Thuan Van Tran [1,2], Vinh Huu Nguyen [1,2], Long Giang Bach [2], Nhan Viet Tran [3], Dai-Viet N. Vo [1], Tuyen Van Nguyen [4], Seong-Soo Hong [5] and Sy Trung Do [4]

1. Center of Excellence for Green Energy and Environmental Nanomaterials (CE@GrEEN), Nguyen Tat Thanh University, 300A Nguyen Tat Thanh, Ho Chi Minh City 755414, Vietnam; kimoanhnguyen88@gmail.com (O.K.T.N.); tranuv@gmail.com (T.V.T.); nguyenhuuvinh3110@gmail.com (V.H.N.); vo.nguyen.dai.viet@gmail.com (D.-V.N.V.)
2. NTT Institute of High Technology, Nguyen Tat Thanh University, 300A Nguyen Tat Thanh Street, Ho Chi Minh City 755414, Vietnam; blgiang@ntt.edu.vn
3. Faculty of Chemical Technology, Ho Chi Minh City University of Food Industry, 140 Le Trong Tan, Ho Chi Minh City 705800, Vietnam; nhantv25101997@gmail.com
4. Institute of Chemistry, Vietnam Academy of Science and Technology, 18 Hoang Quoc Viet, Hanoi 10072, Vietnam; ngvtuyen@hotmail.com (T.V.N.); dosyvhh@gmail.com (S.T.D.)
5. Department of Chemical Engineering, Pukyong National University, 365 Shinsunro, Nam-ku, Busan 48547, Korea; sshong@pknu.ac.kr
* Correspondence: ndtrinh@ntt.edu.vn; Tel.: +84-971-275-356

Received: 28 June 2019; Accepted: 26 August 2019; Published: 1 November 2019

Abstract: A bimetallic metal–organic framework material, which was generated by bridging iron (III) cations and nickel (II) cations with 1,4-Benzenedicarboxylic anions (Fe$_2$Ni-BDC), was synthesized by a solvothermal approach using nickel (II) nitrate hexahydrate and iron (III) chloride hexahydrate as the mixed metal source and 1,4-Benzenedicarboxylic acid (H$_2$BDC) as the organic ligand source. The structure of samples was determined by X-ray powder diffraction (XRD), Fourier transform infrared spectroscopy (FT-IR), Raman spectroscopy, and nitrogen physisorption measurements. The catalytic activity and recyclability of the Fe$_2$Ni-BDC catalyst for the Michael addition amidation reaction of 2-aminopyridine and nitroolefins were estimated. The results illustrated that the Fe$_2$Ni-BDC catalyst demonstrated good efficiency in the reaction under optimal conditions. Based on these results, a reaction mechanism was proposed. When the molar ratio of 2-aminopyridine and trans-β-nitrostyrene was 1:1, and the solvent was dichloromethane, the isolated yield of pyridyl benzamide reached 82%; at 80 °C over 24 h. The catalyst can be reused without a substantial reduction in catalytic activity with 77% yield after six times of reuse.

Keywords: metal–organic framework; bimetallic metal–organic frameworks; decarboxylative amidation

1. Introduction

The amide bonds existing in a large number of structures and forming the backbone of the biologically essential proteins are the most basic building blocks of chemistry in nature [1]. The amide is also essential due to its role in the peptide bonds in pharmaceuticals, proteins, and natural products [1]. Amide bonds are characteristically synthesized by combining carboxylic acids and amines; however, the association of these two functional groups does not occur at room temperature [2]. Over the past decades, many researchers have proposed alternative synthesis methods, such as the Staudinger reaction [3], direct amidation of aldehydes [4], transition-metal-catalyzed aminocarbonylation [5], and hydrating coupling of alkynes with azides [6], esters [7], alcohols [8], and alkynes with amines [9]. Pyridyl benzamides are one of the most critical types of N-heterocyclic amides and play an essential role in the composition of many important medicines (e.g., antiulcer agents, kinetoplastid inhibitors, antifungal agents, and luciferase inhibitors) [10]. Lately, the number of methods for the synthesis of pyridyl benzamides has been rising significantly. Typical examples include the Cu-catalyzed oxidation of methyl ketones [11], Cu-catalyzed oxidative coupling of 2-aminopyridines and terminal alkynes with visible light mediation [12], and the direct oxidative amidation reaction of aldehydes with amines [13]. Since N-heterocyclic amides are increasingly being used in medicine, the discovery of a more effective approach to synthesize pyridyl benzamides is of great importance in the medical industry and has thus been attempted via numerous routes. Xiao-Lan Xu et al. synthesized N-pyridinyl benzamide from benzoylformic acid and aniline using a transition metal catalyst, AgOTf [14]. Additionally, Leiling Deng et al. also created N-pyridinyl benzamide from 2-aminopyridine and phenylacetic acid in the presence of a Cu salt catalyst [15]. Very recently, Zhengwang Chen et al. performed a reaction to synthesize N-pyridinyl benzamide from 2-aminopyridine and *trans*-β-nitrostyrene by utilizing $Ce(NO_3)_3 \cdot 6H_2O$ catalyst in the absence of any oxidant or additive [16].

Metal–organic frameworks (MOFs), a class of porous materials with excellent potential, have been increasingly used in gas storage and separation because of their high capacity and selectivity properties [17]. MOF crystals are built through the formation of secondary building units (SBUs) consisting of organic linkers and metal ions/clusters. Numerous MOF structures have been designed to obtain different features such as enlarged surface areas [18], enhanced catalytic activity and electrical conductivity [19], better interaction at the open metal sites [20], and improved adsorption [21]. Notably, the application of MOFs as effective catalysts for organic reactions has received much attention [22]. MOFs with metal nodes (metal ions/clusters) can act as the active catalytic sites for many organic reactions such as oxidation reactions, C–C coupling reactions, and hydrogenation reactions. However, the fabrication of a MOF structure with high activity and selectivity still requires further investigation.

Recently, a new bimetallic metal–organic framework (BMOF) with a synergistic effect between different metal ions was developed. Different from the synthesis of MOFs where only one metal ion is combined with organic ligands, the synthesis process of a BMOF produces the pure phase of the BMOF by combining two different metal ions with organic ligands [23]. The BMOFs are expected to have improved stability, activity, and surface area and could be applicable in catalysis. For example, an Fe/Co mixed Hofmann MOF with coupling effects between Co^{2+} and Fe^{2+} ions was found to exhibit enhanced catalytic activity for an oxygen evolution reaction compared to that exhibited by the original single-metal MOFs [24]. Despite this, the catalysis applications of BMOFs have not been investigated so far.

The main aim of this study was to appraise the effect of a new BMOF (Fe_2Ni-1,4-Benzenedicarboxylic, Fe_2Ni-BDC) as a heterogenous catalyst for organic synthesis reactions. Fe_2Ni-BDC was generated by bridging iron (III) cations and nickel (II) cations with 1,4-Benzenedicarboxylic anions (BDC^-) to create a porous three-dimensional structure [25,26]. We also investigated the synthesis of N-pyridinyl benzamide via the amidation process of *trans*-β-nitrostyrene and 2-aminopyridine using Fe_2Ni-BDC as an effective heterogeneous catalyst, without using added reducing agents or oxidizing agents. This catalyst might be reused for the creation of N-pyridinyl benzamide by the amidation reaction without significant depreciation in its efficiency. Fe_2Ni-BDC is also satisfactory from the view of green chemistry, as the solid catalyst used for the reaction can be easily recovered and reused. To the best of our knowledge, C=C double bond cleavage has not previously been performed using heterogeneous catalysis.

2. Experimental

2.1. Synthesis of the Catalyst

The bimetallic Fe/Ni-BDC catalyst was synthesized by the solvothermal method [27–29]. Typically, a clear solution containing H$_2$BDC (9 mmol, Sigma-Aldrich, Saint Louis, MO, USA), FeCl$_3$·6H$_2$O (6 mmol, Fisher Scientific, Hampton, NH, USA), and Ni(NO$_3$)$_2$·6H$_2$O (1.8 mmol, Fisher Scientific, Hampton, NH, USA) in 60 mL of N,N-dimethylformamide (DMF, 99%, Xilong Chemical Co., Ltd., Guangzhou, China) was prepared and placed in a pressure device (100 mL Hydrothermal Synthesis Autoclave Reactor 304 Stainless Steel High-Pressure Digestion Tank with PTFE Lining for Rapid Digestion of Insoluble Material, Baoshishan, Shanghai, China). Then, this device was placed in an oven (Memmert UN110, Schwabach, Germany) at 100 °C for three days. The non-reacted components, such as the remains of the organic linker in porous holes was elimated by a distillation router with DMF solvent at 100 °C for a day. The orange solid was washed in DMF solvent (three times) and water (three times), followed by drying for a day at 60 °C. For comparison, Ni-BDC and Fe-BDC catalysts were also synthesized via the same method as for the synthesis of Fe$_2$Ni-BDC by using Ni(NO$_3$)$_2$·6H$_2$O for the synthesis of Ni-BDC and FeCl$_3$·6H$_2$O for the synthesis of Fe-BDC. The catalyst products obtained were an orange solid and a green solid for the Fe-BDC and Ni-BDC materials, respectively.

2.2. Catalyst Characterization

Physical and chemical methods are used to determine the characteristic properties of MOFs. X-ray diffraction (XRD) analysis was employed on a Bruker AXS D8 Advantage (Bruker, Billerica, MA, USA) operating with a Cu Kα source to determine the material structure. The specific surface area and pore distribution of the obatined catalysts were determined using a Nova Quantachrome 2200e (Quantachrome Instruments, Kingsville, TX, USA). Samples were activated in a vacuum at 150 °C for 6 h, followed by nitrogen adsorption at 77 °C and low pressure. Thermogravimetric analysis (TGA) was conducted on a Netzsch Thermoanalyzer STA 409 (Netzsch, Selb, Germany) with a heating rate of 10 °C/min from room temperature to 800 °C under inert gas conditions. The infrared spectrum (FT-IR) was attained using a Bruker TENSOR37 (Bruker, Billerica, MA, USA), and a KBr compressed sample was used to determine functional groups in the material. SEM images of the catalyst were obtained using a scanning electron microscope (SEM) on a JSM 7401F device (Jeol, Peabody, MA, USA).

2.3. The Synthesis of N-Pyridinyl Benzamide

In a typical catalytic experiment, trans-β-nitrostyrene (**1a**, 0.2 mmol, 0.0298 g) and 2-aminopyridine (**2a**, 0.2 mmol, 0.0188 g) in dichloromethane (DCM) solvent (1 mL) in the presence of Fe$_2$Ni-BDC were added into the pressure equipment. The reaction mixture was stirred at 80 °C for 24 h in atmospheric air (Table S1). Following this stage, the compound was cooled down to room temperature. The anticipated products were isolated using column chromatography. GC-MS, ^{1}H NMR, and ^{13}C NMR analyses were employed to determine the product structure (Supplementary Materials). To assess the recovery and reusability of the catalyst, the catalyst was separated and washed thoroughly with large amounts of ethanol, dried at 100 °C, and reused for further experiments.

3. Results and Discussion

3.1. Characterization of the Ni-BDC, Fe-BDC, and Fe$_2$Ni-BDC Catalysts

In this study, a BMOF based on the coupling effects between Ni^{2+} and Fe^{2+} ions (Fe$_2$Ni-BDC) and the respective single-metal-ion MOFs (Ni-BDC and Fe-BDC) were obtained by direct synthesis with a clear solution containing nickel (II) nitrate hexahydrate and/or iron (III) chloride hexahydrate and terephthalic acid (H$_2$BDC) in dimethylformamide (DMF) solvent. Evidence of the formation of a MOF with bimetallic nodes was confirmed by X-ray powder diffraction (XRD), Fourier transform infrared spectroscopy (FT-IR), Raman spectroscopy, and nitrogen physisorption measurements.

The XRD patterns of the resulting Ni-BDC, Fe-BDC, and Fe$_2$Ni-BDC samples are shown in Figure S3 (Supplementary Materials). The pure Ni-BDC powder exhibited a similar XRD pattern (Figure S3a) to the previously reported ones synthesized by a solvothermal method with main diffraction peaks at 2θ of 11°, 11.5°, 14°, 15°, 16.5°, 17,5°, 28°, and 29° [27,28,30–32]. As can be observed in Figure 1b, the XRD patterns of the Fe-BDC exhibited peaks at 2θ of approximately of 9.2°, 9.5°, 14.0°, 16.4°, and 18.7°, and this result was also similar to the simulated patterns of MIL-53(Fe) previously reported in the literature [27,33]. Furthermore, the simulated diffraction patterns for the Ni- and Fe-based system (see Figures S1 and S2, Supplementary Materials) were exhibited. In the pattern of the Fe$_2$Ni-BDC samples (Figure S3c, Supplementary Materials), the XRD peaks emerged around 2θ of 7.4°, 8.8°, 9.2°, 9.8°, 16.7°, 18.7°, 17.8°, 20.0°, and 21.8°. It is clear that this result is also in line with those of the sample previously reported in the literature [25,26]. Besides this, no other diffraction peak combined with iron oxides, nickel oxides, or other diffraction peaks was found, proving the high purity of the samples. The XRD analytic results illustrate that the crystal structure of Fe$_2$Ni-BDC was substantially affected by the existence of assorted metal ions (Ni^{2+} and Fe^{3+} ions), and Fe$_2$Ni-BDC crystals could be formed by the combination of H$_2$BDC with Ni^{2+} and Fe^{3+} ions [25,26].

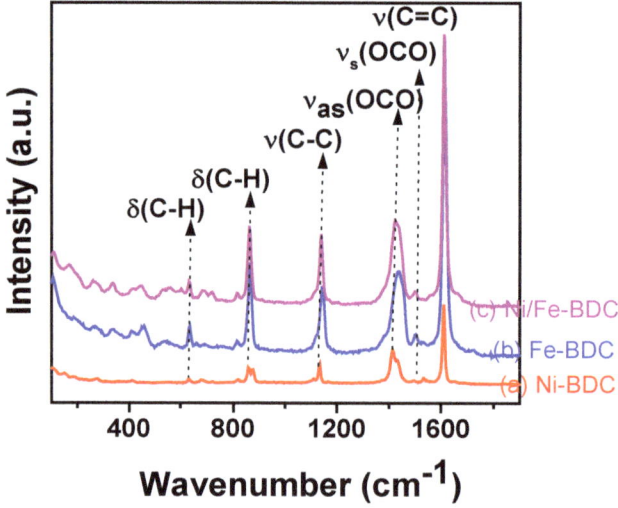

Figure 1. Raman spectroscopy of the Ni-1,4-Benzenedicarboxylic (Ni-BDC) (**a**), Fe-1,4-Benzenedicarboxylic (Fe-BDC) (**b**), and Fe$_2$Ni-1,4-Benzenedicarboxylic (Fe$_2$Ni-BDC) (**c**).

The FT-IR spectra of Ni-BDC, Fe-BDC, and Fe$_2$Ni-BDC are shown in Figure S4 (Supplementary Materials). The FT-IR spectra exhibited stretching vibration of the C=O bond at approximately 1680 cm^{-1}, while υ_{asym} (OCO) and υ_{sym} (OCO) bonds displayed stretching vibration at around 1601 cm^{-1} and 1391 cm^{-1}, respectively. Besides this, the FT-IR spectra displayed stretching vibration of the υ(C–O) and δ(C–H) bonds at 1017 cm^{-1} and 749 cm^{-1} (Figure S4A, Supplementary Materials). These results showed that the existence of the metal–ligand bond in the MOF structures. Particularly, the FT-IR spectrum of H$_2$BDC displayed no band at 1700 cm^{-1}. This result proves the absence of free H$_2$BDC in the MOF structures [34,35]. The feature bands of H$_2$O and DMF in the MOF materials were exhibited at 1657 cm^{-1} and 3387 cm^{-1} [25,26]. At lower frequencies, stretching vibration of the C–H bond, C=C bond, and –OCO function was observed at approximately 750 cm^{-1}, 690 cm^{-1}, and 660 cm^{-1}, respectively, proving the existence of the vibrations of the organic linker BDC (Figure S4B, Supplementary Materials) [26]. Moreover, it is clear that the strong band at 547 cm^{-1} may be attributed to either NiO vibrations or FeO vibrations [36]. The weak range at about 720 cm^{-1} is associated with

Fe$_2$NiO vibration, which was also detected in the Fe$_2$Ni-BDC sample [26]. This result reinforces the notion that Ni^{2+} and Fe^{3+} ions may combine with H$_2$BDC to form Fe$_2$Ni-BDC crystals.

The Raman spectroscopy results of Ni-BDC, Fe-BDC, and Fe$_2$Ni-BDC are shown in Figure 1. Following previous studies, the symmetric oscillation modes and asymmetric oscillation of the COO– bond in the carboxylate group detected at approximately 1445 cm^{-1} and 1501 cm^{-1} may be the organic linker BDC in the metal–organic frameworks. The oscillation at around 1140 cm^{-1} can be attributed to the C–C bond of the carboxylate group with a benzene ring. Besides this, vibration of the C–H bond was observed at around 865 cm^{-1} and 630 cm^{-1} [25]. As shown in Figure 1, the presence of BDC ligand was also discovered in the catalyst sample, and no Raman sign corresponding to NiO, FeO, or other impurities was detected on the pattern, which is consistent with the X-ray diffraction results.

Concurrent thermal analysis permits simultaneous measurement of both the weight and heat flow alteration of Ni-BDC, Fe-BDC, and Fe$_2$Ni-BDC powder in relation to the temperature under an air atmosphere. The patterns show the differential scanning calorimetry (DSC) and thermogravimetry (TGA) curved of Ni-BDC, Fe-BDC, and Fe$_2$Ni-BDC powder from room temperature to 800 °C under an air atmosphere (Figure 2). In the DSC curve of Ni-BDC, a powerful exothermic process occurred between 380 °C and 480 °C, manifesting as a peak temperature at 450.57 °C (Figure 2a). In the DSC curve of Fe-BDC, a robust exothermic process occurred between 260 °C and 340 °C, illustrating a peak temperature at 316.53 °C (Figure 2b). In the DSC curve of Fe$_2$Ni-BDC, a strongly exothermic process took place from 320 °C to 490 °C, signified by a peak temperature at 437.99 °C (Figure 2c). The weight loss occurring at temperatures below 200 °C could be attributable to the vaporization of solvent (H$_2$O or DMF) obstructed within the frame, while the weight loss occurring between 200 °C and 260 °C is the result of the strong combining of H$_2$O or the frame of H$_2$O. A sudden weight loss may be discerned between 280 and 480 °C, conforming to a strongly exothermic process in the DSC curve. Herein, elemental analysis, i.e., EDX mapping and ICP analysis, can aid the identification of the Ni/Fe ratio in the bimetallic Ni/Fe-BDC MOF. Based on the results obtained from EDX mapping, we found that the proximate percentages of Ni and Fe were 5.7% and 11.2%, respectively (Figure S5, Supplementary Materials) or 1:2 when calculated in a molar ratio. Similarly, ICP analysis also indicated that the molar (atomic) ratio between Ni and Fe was at approximately 1:2. Therefore, it matched well with the formula of NiFe$_2$-BDC.

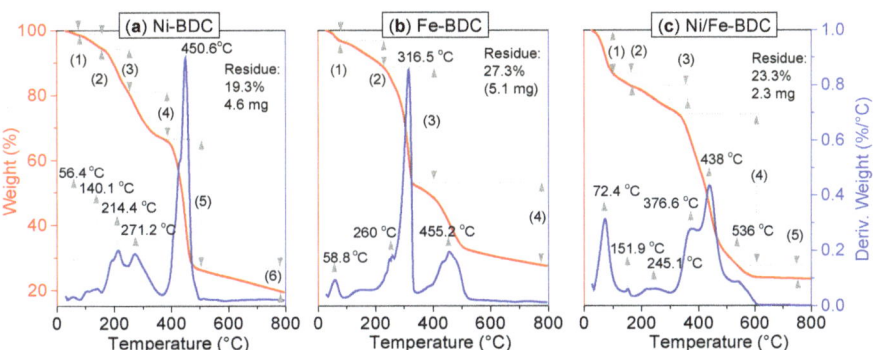

Figure 2. TGA analysis of the Ni-1,4-Benzenedicarboxylic (Ni-BDC) (**a**), Fe-1,4-Benzenedicarboxylic (Fe-BDC) (**b**), and Fe2Ni-1,4-Benzenedicarboxylic (Fe2Ni-BDC) (**c**).

The morphology, size, and regularity of the Ni-BDC, Fe-BDC, and Fe$_2$Ni-BDC samples were studied by SEM (Figure 3). The Ni-BDC sample includes stacked planar sheets with a size up to several micrometers (Figure 3a). The shape of pure Fe-BDC showed two types of particles: bigger, rod-like particles and other, smaller, pseudo-spherical particles (Figure 3b). The SEM pictures of the particles unveiled the creation of uniform micro-sized hexagonal rods (Figure 3c). Besides this, the Fe$_2$Ni-BDC sample consists of stacked planar sheets and other smaller pseudo-spherical particles.

Figure 3. SEM images of Ni-1,4-Benzenedicarboxylic (Ni-BDC) (**a**), Fe-1,4-Benzenedicarboxylic (Fe-BDC) (**b**), and Fe2Ni-1,4-Benzenedicarboxylic (Fe2Ni-BDC) (**c**).

The surface areas of the catalysts were confirmed by nitrogen adsorption–desorption isotherms derived from Brunauer–Emmett–Teller (BET). The isotherms of Ni-BDC, Fe-BDC, and Fe$_2$Ni-BDC are presented in Figure 4. The BET surface areas of Fe-BDC and Fe$_2$Ni-BDC were 158 m^2/g and 247 m^2/g, respectively. Meanwhile, the BET surface area of Ni-BDC was extremely low at around 2.28 m^2/g, and it does not seem to be porous. The mesopore size distribution curves of specimens calculated using the Barrett–Joyner–Halenda (BJH) model are displayed in Figure 4. The pore volume and pore width of Ni-BDC, Fe-BDC, and Fe$_2$Ni-BDC suggested average pore sizes of about 25 nm, 11 nm, and 13 nm, respectively. Based on the above results, including XRD, FT-IR, Raman, DSC, TGA, and BET, we conclude that an Fe$_2$Ni-BDC bimetallic metal–organic framework was successfully synthesized by the solvothermal approach.

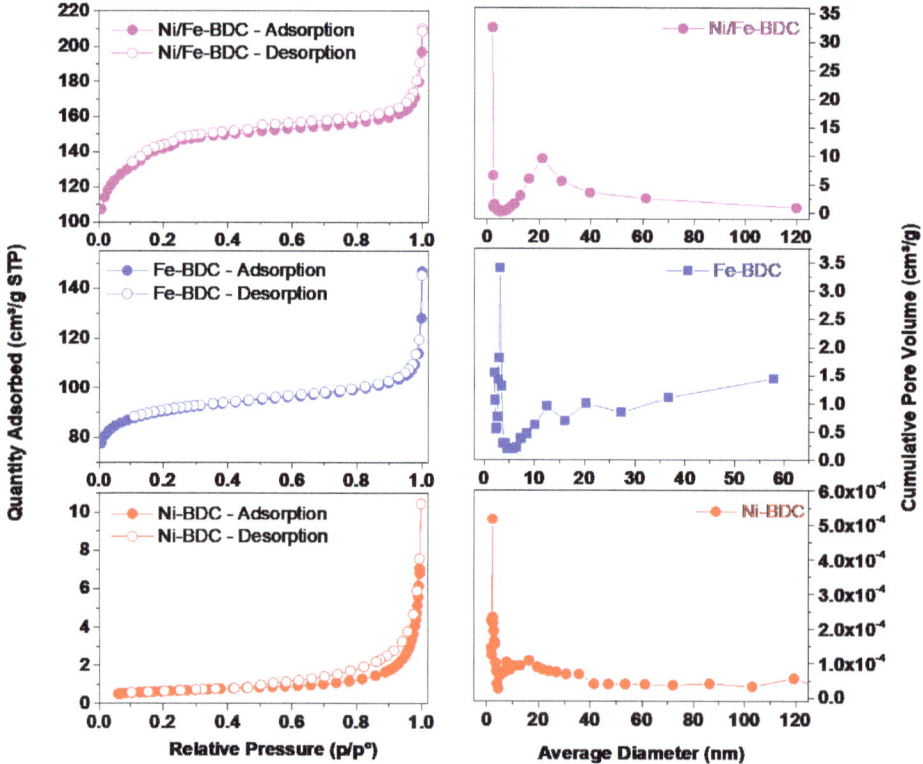

Figure 4. N$_2$ adsorption–desorption isotherms (left) and pore size distributions (right) of metal–organic framework (MOF) samples.

3.2. The Synthesis of N-Pyridinyl Benzamide

Scheme 1 illustrates the amidation reaction between trans-β-nitrostyrene and 2-aminopyridine using catalysts $Ni(NO_3)_2 \cdot 6H_2O$, $FeCl_3 \cdot 6H_2O$, Ni-BDC, Fe-BDC, and Fe_2Ni-BDC. The performance of the reactions with different metal-centered catalysts demonstrated that Fe_2Ni-BDC resulted in the best activity for this amidation process (Figure 5a, Table S1).

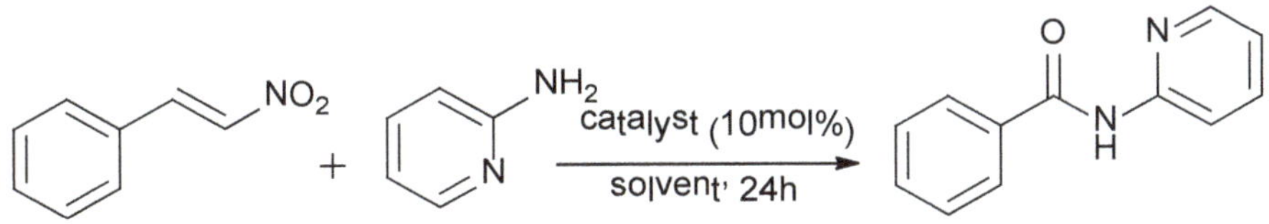

Scheme 1. The amidation reaction between trans-β-nitrostyrene and 2-aminopyridine using Fe_2Ni-BDC as a catalyst.

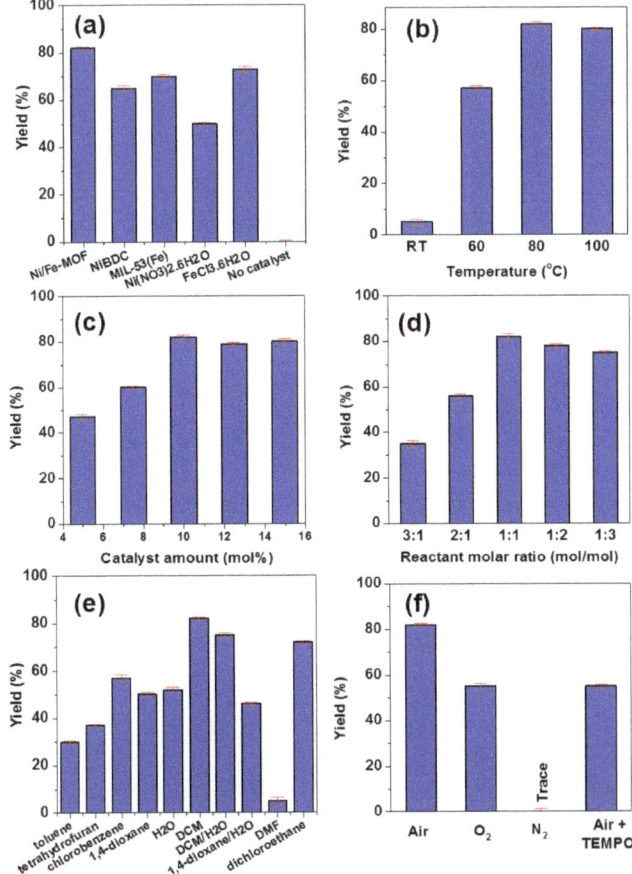

Figure 5. The yield of N-pyridinyl benzamide vs. catalyst (**a**), temperature (**b**), catalyst amount (**c**), reaction molar ratio (**d**), solvents (**e**), and yield of N-pyridinyl benzamide under air, O_2, N_2, and air + 2,2,6,6-Tetramethyl-1-piperidinyloxy (TEMPO) (**f**).

The amidation reaction was performed at different temperatures from room temperature to 100 °C. The results (Figure 5b and Table S1) indicated that lower temperature led to a decline in product yield, and the reaction occurred when the mixture was heated. Also, when this reaction was carried out at 100 °C, a lower yield of product **3a** was observed, which could result from the decomposition of reactants and products, adversely affecting the reaction process. The yield of product **3a** reached its highest value when the reaction was carried out at 80 °C; therefore, 80 °C was chosen as the optimal reaction temperature for further studies.

Figure 5c shows the effect of catalyst amount (mol %) on the yield of product **3a**. These results revealed that the yield of **3a** reached a peak (82%) when using 10 mol % catalyst. This behavior necessitates the bimetallic framework for catalyzing the alteration. The use of BMOF catalyst for the reaction brought about a remarkable progression in the reaction yield. The reaction utilizing 5 mol % of catalyst might achieve around 47% yield, and this yield figure could be improved to 60% yield if 7.5 mol % catalyst is used. However, the use of more than 10 mol % catalyst seemed to be redundant because the yield of **3a** was not improved substantially above this concentration (Table S1).

We also investigated the effect of the trans-β-nitrostyrene/2-aminopyridine molar ratio on the production of N-(pyridin-2-yl)-benzamide. The reaction was carried at 80 °C under air in DCM for 24 h in the presence of 10 mol % of catalyst. The survey data illustrated that the reactant molar ratio exerted a significant effect on the yield of product **3a**. The reaction utilizing the proportion of 1 equivalent of **1a** achieved 82% yield in air. The excess **1a** reactant resulted in a reduced yield of **3a**. Indeed, as the yield of the product reached 56%, the molar proportion of reactants in the reaction approximated 2:1. However, the yield was decreased drastically to merely 35% when 3 equivalents of **1a** were used. It was clear that excess **2a** was preferred in the reaction. The yield of product **3a** reached approximately 78% after 24 h with a reactant molar proportion of either 1:2 or 1:3. However, the optimal yield of **3a** was attained at the reactant molar proportion of 1:1 (Figure 5d, Table S1).

Because the amidation reaction of **1a** and **2a** was carried out in the liquid phase, we needed to examine the effect of solvents on the catalytic activity. In the first report of the synthesis of N-(pyridin-2-yl)-benzamide derivatives from **1a** and **2a**, Zhengwang Chen et al. [16] implemented the reaction in various co-solvents and illustrated that co-solvent H_2O/dioxane could improve the yield of products. The effect of solvents such as toluene, DMF, 1,4-dioxane, tetrahydrofuran (THF), chlorobenzene, DCM/H_2O, dioxane/H_2O, H_2O, and dichloroethane on the yield of **3a** was investigated. As shown in Figure 5e and Table S1, DMF solvent was unsuitable for the reaction utilizing the bimetallic framework catalyst, with only 5% yield of **3a** after 24 h. The yields of product **3a** were higher in toluene and tetrahydrofuran—30% and 37% yields, respectively, after 24 h. The reactions controlled in dioxane, dioxane/H_2O, H_2O, and chlorobenzene produced **3a** in yields of 52%, 46%, 50%, and 57%, respectively. Higher yields of 75% and 72% were obtained for DCM/H_2O and dichloroethane. Among the aforementioned solvents, DCM gave the best result with an 82% yield of **3a** after 24h (Figure 5e and Table S1).

As shown in Figure 5f, the production of **3a** was not detected when the reaction was conducted under N_2 gas, highlighting the essential role of O_2 in the reaction. Further investigation of the reaction with the bimetallic framework catalyst under O_2 illustrating the role of the Fe_2Ni-BDC catalyst was not feasible. Besides this, the reaction that was carried out in the presence of 2,2,6,6-Tetramethyl-1-piperidinyloxy (TEMPO) revealed no obvious prevention of the reaction, allowing us to propose a probably easier radical path in this amidation reaction (Figure 5f, Table S1). After some surveys, we determined that the appropriate reaction conditions are as follows: trans-β-nitrostyrene (**1a**, 0.2 mmol), 2-aminopyridine (**2a**, 0.2 mmol), catalyst (10 mol %), and DCM (1 mL) at 80 °C for 24 h.

A leaching evaluation was employed to confirm whether the active sites going into solution from the solid catalyst could accelerate the creation of N-(pyridin-2-yl)-benzamide, as the leaching phenomenon could happen throughout the stages of the reaction. The reaction was performed at 80 °C in DCM solvent for 24 h, with reactants consisting of trans-1-nitro-phenylethylene and 2-pyridyl amine and with 10 mol % catalyst under air atmosphere. After the initial 4 h stage with 12% yield recorded, the catalyst was separated by centrifugation. Then, new reactants were added to the solution phase in a clean, pressurized vial with magnetic stirring and the temperature maintained at 80 °C for 24 h. It was observed that the formation of N-(pyridin-2-yl)-benzamide stopped after the catalyst was separated (Figure 6a). These figures indicate that the amidation of *trans*-1-nitro-phenylethylene with 2-aminopyridine to generate N-(pyridin-2-yl)-benzamide could be maintained only in the presence of the solid catalyst.

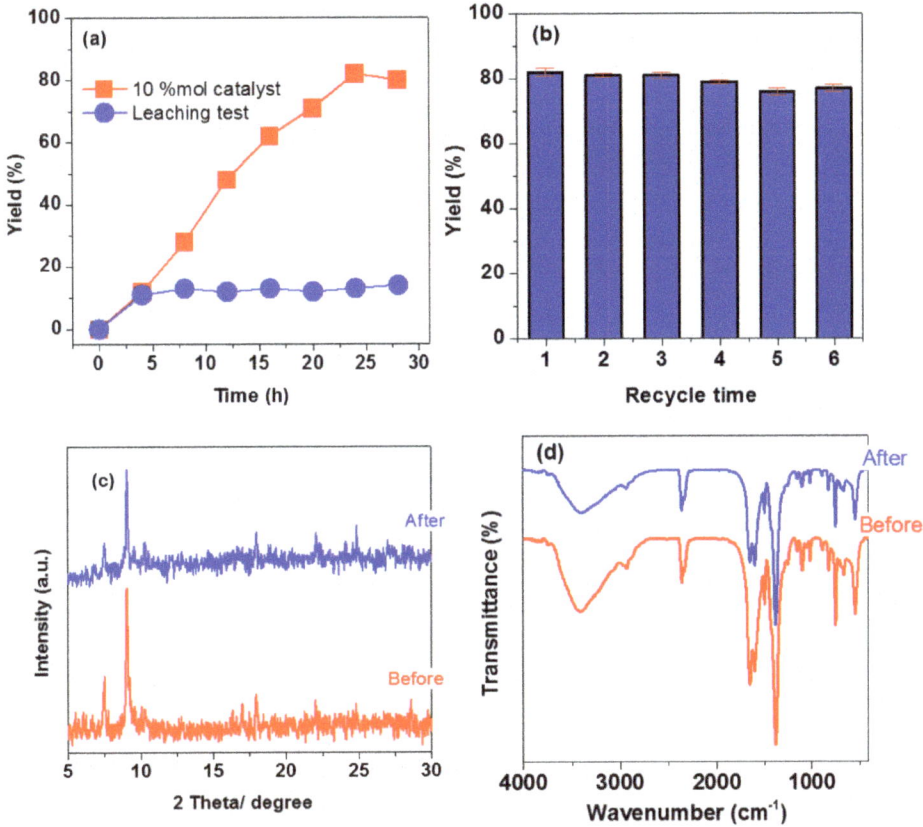

Figure 6. Leaching test (**a**), catalyst reusability (**b**), XRD pattern (**c**), and FT-IR spectroscopy (**d**) of the fresh and recovered catalyst.

Another striking feature that differentiates heterogeneous catalysis from homogeneous catalysis is the capability of the catalyst to be recovered and recycled. The catalyst was appropriately surveyed for recyclability in the reaction over six successive runs. The reaction was performed under optimal conditions at 80 °C in an air atmosphere. Upon completion of the first run, the catalyst was separated, washed cautiously with DCM and DMF, and dried at 100 °C for 3 h. Afterward, the recovered catalyst was recycled in new experiments. After the experiments were conducted, these data demonstrated that the catalyst might be recycled numerous times in the reaction of 2-aminopyridine and *trans*-1-nitro-phenylethylene to create N-(pyridin-2-yl)-benzamide without significantly compromising the yield. The yield of N-(pyridin-2-yl)-benzamide noted in the sixth run was 77% (Figure 6b). Moreover, the whole catalyst might be withheld after the reaction process, as shown by XRD (Figure 6c) and FT-IR (Figure 6d) spectroscopy of the recovered BMOFs.

Based on the experimental results, a reasonable mechanism is suggested in Scheme 2. Initially, the intermediary **A** was created via the Michael addition of *trans*-β-nitrostyrene **1a** and 2-pyridine amines **2a** as a nucleophile. When the catalyst played the role of a Lewis acid, the intermediary **A** was able to form a bond covalent with the O molecule on the nitro group, permitting the Michael addition, because the Fe^{3+} and Ni^{2+} ions in the BMOFs had many empty orbitals in the molecule. Intermediary **C** formed through successive dehydration of **B** and was reorganized to imine intermediary **D**. Following that, hydration of **E** brought about the intermediary **F**. After that, protonation of **F** took place by dehydration to provide the α-aminonitrile to intermediary **G**. Eventually, the target product **3a** was formed via a nucleophilic addition and elimination process. Via this proposed mechanism, the catalyst with two active metal centers improved the reaction yield. Therefore, the catalysis activity of Fe_2Ni-BDC was enhanced when compared with the catalysis activity of the single-metal centers of Ni-BDC and Fe-BDC.

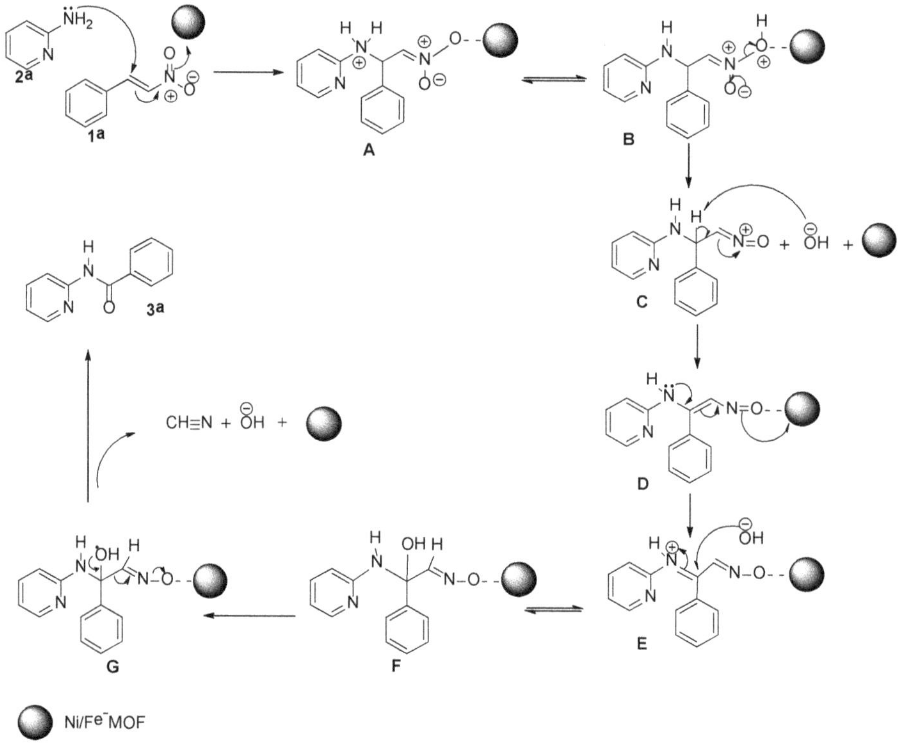

Scheme 2. Proposed reaction mechanism.

The study was subsequently extended to the synthesis of various N-(pyridin-2-yl)-benzamide derivatives. The reactions were conducted between derivatives of 2-aminopyridine and *trans*-β-nitrostyrene in DCM solvent for 24 h under air atmosphere at 80 °C with 10 mol % of the catalyst. The products were purified by column chromatography, and isolated yields were noted. As shown in Table 1, N-(pyridin-2-yl)-benzamide was created with 82% yield (entry 1, Table 1, Figures S6 and S7). The existence of a substituent on the pyridine ring in 2-aminopyridine reduced the yield slightly. The reaction performed between 4-methyl-2-aminopyridine and *trans*-β-nitrostyrene produced N-(4-methylpyridin-2-yl)benzamide with 78% yield (entry 2, Table 1, Figures S8 and S9), while a 68% yield of N-(5-chloropyridin-2-yl)benzamide was obtained for a reaction between 5-cloro-2-aminopyridine and *trans*-β-nitrostyrene (entry 3, Table 1, Figures S10 and S11). Besides these, the reaction conducted between 4-methyl-o-phenylenediamine and *trans*-β-nitrostyrene produced 5-methyl-2-phenyl-1*H*-benzo[*d*]imidazole with 63% yield (entry 4, Table 1, Figures S12 and S13). Finally, N-(pyridin-2-yl)-benzamide was still generated when the reaction was performed between benzoylformic acid and 2-aminopyridine with 74% yield (entry 5, Table 1, Figures S14 and S15).

Table 1. Synthesis of different N-(pyridin-2-yl)-benzamide derivatives utilizing Fe_2Ni-BDC catalyst.

Entry	Reactant 1	Reactant 2	Product	Isolated Yields (%)
1	PhCH=CH-NO$_2$	2-aminopyridine	N-(pyridin-2-yl)-benzamide	82 (85 [16])
2	PhCH=CH-NO$_2$	4-methyl-2-aminopyridine	4-methyl derivative	78
3	PhCH=CH-NO$_2$	5-chloro-2-aminopyridine	5-chloro derivative	68
4	PhCH=CH-NO$_2$	4-methyl-o-phenylenediamine	5-methyl-2-phenyl-1H-benzimidazole	63
5	benzoylformic acid (PhCOCOOH)	2-aminopyridine	N-(pyridin-2-yl)-benzamide	74

4. Conclusions

The bimetallic metal–organic framework Fe_2Ni-BDC is a productive heterogeneous catalyst for the amidation reaction between *trans*-1-nitro-phenylethylene and 2-aminopyridine to create N-(pyridin-2-yl)-benzamide under air. Fe_2Ni-BDC showed higher productivity in the synthesis of N-(pyridin-2-yl)-benzamide than other metal–organic frameworks. The bimetallic metal–organic framework was surveyed as a heterogeneous catalyst for the amidation reaction. The catalyst was successfully recovered and reused for the reaction generating N-(pyridin-2-yl)-benzamide without a reduction in catalyst activity. To the best of our knowledge, the formation of N-(pyridin-2-yl)-benzamide has not been previously achieved utilizing a heterogeneous catalyst.

Supplementary Materials: The following are available online at http://www.mdpi.com/2227-9717/7/11/789/s1: 1. General experimental information. 2. General procedure for the synthesis of N-pyridyl benzamide. 3. General procedure of investigation for the synthesis of N-pyridinyl benzamide. 4. Characterization data for all products. Table S1: Optimization of reaction conditions. Figure S1: The presented simulated diffraction patterns for Ni-based was based on the corresponding check CIF file of Ni-BDC compare with experimental patterns. Figure S2: The presented simulated diffraction patterns for Fe-based was based on the corresponding check CIF file of MIL-53 (Fe) compare with experimental patterns. Figure S3: X-ray powder diffraction of Ni-BDC, Fe-BDC and Fe_2Ni-BDC. Figure S4: FT-IR spectra of the Ni-BDC, Fe-BDC, and Fe2Ni-BDC. Figure S5: EDX mapping point of Fe_2Ni-BDC. Figure S6: ^{1}H-NMR spectra of N-(pyridin-2-yl)benzamide. Figure S7: ^{13}C-NMR spectra of N-(pyridin-2-yl)benzamide.

Figure S8: ^{1}H-NMR spectra of N-(4-methylpyridin-2-yl)benzamide. Figure S9: ^{13}C-NMR spectra of N-(4-methylpyridin-2-yl)benzamide. Figure S10: ^{1}H-NMR spectra of 5-methyl-2-phenyl-1H-benzo[d]imidazole. Figure S11: ^{13}C-NMR spectra of 5-methyl-2-phenyl- 1H-benzo[d]imidazole. Figure S12: ^{1}H-NMR spectra of N-(5-chloropyridin-2-yl)benzamide. Figure S13: ^{13}C-NMR spectra of N-(5-chloropyridin-2-yl)benzamide. Figure S14: 1H-NMR spectra of N-(pyridin-2-yl)benzamide. Figure S15: ^{13}C-NMR spectra of N-(pyridin-2-yl)benzamide.

Author Contributions: Data curation, O.K.T.N., V.H.N. and N.V.T.; Formal analysis, T.V.T. and S.T.D.; Methodology, T.V.T., V.H.N., N.V.T. and T.V.N.; Supervision, T.D.N.; Writing—original draft, O.K.T.N.; Writing—review and editing, L.G.B., D-.V.N.V., T.V.N., S.-S.H. and S.T.D.

Funding: This work was supported by the Vietnam National Foundation for Science and Technology Development (NAFOSTED) under grant number 104.01-2019.16.

Conflicts of Interest: The authors declare no conflict of interest.

References

1. Lundberg, H.; Tinnis, F.; Selander, N.; Adolfsson, H. Catalytic amide formation from non-activated carboxylic acids and amines. *Chem. Soc. Rev.* **2014**, *43*, 2714–2742. [CrossRef] [PubMed]
2. Valeur, E.; Bradley, M. Amide bond formation: Beyond the myth of coupling reagents. *Chem. Soc. Rev.* **2009**, *38*, 606–631. [CrossRef] [PubMed]
3. Köhn, M.; Breinbauer, R. The staudinger ligation—A gift to chemical biology. *Angew. Chem. Int. Ed.* **2004**, *43*, 3106–3116. [CrossRef] [PubMed]
4. Leow, D. Phenazinium salt-catalyzed aerobic oxidative amidation of aromatic aldehydes. *Org. Lett.* **2014**, *16*, 5812–5815. [CrossRef]
5. Fang, X.; Li, H.; Jackstell, R.; Beller, M. Selective palladium-catalyzed aminocarbonylation of 1,3-dienes: Atom-efficient synthesis of β,γ-unsaturated amides. *J. Am. Chem. Soc.* **2014**, *136*, 16039–16043. [CrossRef]
6. Cassidy, M.P.; Raushel, J.; Fokin, V.V. Practical synthesis of amides from in situ generated copper(I) acetylides and sulfonyl azides. *Angew. Chem. Int. Ed.* **2006**, *45*, 3154–3157. [CrossRef]
7. Morimoto, H.; Fujiwara, R.; Shimizu, Y.; Morisaki, K.; Ohshima, T. Lanthanum(III) triflate catalyzed direct amidation of esters. *Org. Lett.* **2014**, *16*, 2018–2021. [CrossRef]
8. Liu, X.; Jensen, K.F. Multistep synthesis of amides from alcohols and amines in continuous flow microreactor systems using oxygen and urea hydrogen peroxide as oxidants. *Green Chem.* **2013**, *15*, 1538–1541. [CrossRef]
9. Chen, Z.W.; Jiang, H.F.; Pan, X.Y.; He, Z.J. Practical synthesis of amides from alkynyl bromides, amines, and water. *Tetrahedron* **2011**, *67*, 5920–5927. [CrossRef]
10. Ferrins, L.; Gazdik, M.; Rahmani, R.; Varghese, S.; Sykes, M.L.; Jones, A.J.; Avery, V.M.; White, K.L.; Ryan, E.; Charman, S.A.; et al. Pyridyl benzamides as a novel class of potent inhibitors for the kinetoplastid Trypanosoma brucei. *J. Med. Chem.* **2014**, *57*, 6393–6402. [CrossRef]
11. Subramanian, P.; Indu, S.; Kaliappan, K.P. A one-pot copper catalyzed biomimetic route to n -heterocyclic amides from methyl ketones via oxidative c-c bond cleavage. *Org. Lett.* **2014**, *16*, 6212–6215. [CrossRef] [PubMed]
12. Ragupathi, A.; Sagadevan, A.; Lin, C.C.; Hwu, J.R.; Hwang, K.C. Copper(i)-catalysed oxidative C-N coupling of 2-aminopyridine with terminal alkynes featuring a CC bond cleavage promoted by visible light. *Chem. Commun.* **2016**, *52*, 11756–11759. [CrossRef] [PubMed]
13. Patel, O.P.S.; Anand, D.; Maurya, R.K.; Yadav, P.P. Copper-catalyzed highly efficient oxidative amidation of aldehydes with 2-aminopyridines in an aqueous micellar system. *Green Chem.* **2015**, *17*, 3728–3732. [CrossRef]
14. Xu, X.L.; Xu, W.T.; Wu, J.W.; He, J.B.; Xu, H.J. Silver-promoted decarboxylative amidation of α-keto acids with amines. *Org. Biomol. Chem.* **2016**, *14*, 9970–9973. [CrossRef] [PubMed]
15. Deng, L.; Huang, B.; Liu, Y. Copper(II)-mediated, carbon degradation-based amidation of phenylacetic acids toward N-substituted benzamides. *Org. Biomol. Chem.* **2018**, *16*, 1552–1556. [CrossRef] [PubMed]
16. Chen, Z.; Wen, X.; Qian, Y.; Liang, P.; Liu, B.; Ye, M. Ce(III)-catalyzed highly efficient synthesis of pyridyl benzamides from aminopyridines and nitroolefins without external oxidants. *Org. Biomol. Chem.* **2018**, *16*, 1247–1251. [CrossRef]
17. Yang, X.; Xu, Q. Bimetallic metal-organic frameworks for gas storage and separation. *Cryst. Growth Des.* **2017**, *17*, 1450–1455. [CrossRef]

18. Xia, B.Y.; Yan, Y.; Li, N.; Wu, H.B.; Lou, X.W.D.; Wang, X. A metal-organic framework-derived bifunctional oxygen electrocatalyst. *Nat. Energy* **2016**, *1*, 15006. [CrossRef]
19. Chen, W.; Zhang, Z.; Bao, W.; Lai, Y.; Li, J.; Gan, Y.; Wang, J. Hierarchical mesoporous γ-Fe2O3/carbon nanocomposites derived from metal organic frameworks as a cathode electrocatalyst for rechargeable Li-O2 batteries. *Electrochim. Acta* **2014**, *134*, 293–301. [CrossRef]
20. Kim, S.H.; Lee, Y.J.; Kim, D.H.; Lee, Y.J. Bimetallic metal-organic frameworks as efficient cathode catalysts for Li-O2 batteries. *ACS Appl. Mater. Interfaces* **2018**, *10*, 660–667. [CrossRef]
21. Villajos, J.A.; Orcajo, G.; Martos, C.; Botas, J.Á.; Villacañas, J.; Calleja, G. Co/Ni mixed-metal sited MOF-74 material as hydrogen adsorbent. *Int. J. Hydrog. Energy* **2015**, *40*, 5346–5352. [CrossRef]
22. Albero, J.; García, H. Metal organic frameworks as catalysts for organic reactions. In *New Materials for Catalytic Applications*; Elsevier: Amsterdam, The Netherlands, 2016; pp. 13–40. ISBN 9780444635877.
23. Gholipour-Ranjbar, H.; Soleimani, M.; Naderi, H.R. Application of Ni/Co-based metal-organic frameworks (MOFs) as an advanced electrode material for supercapacitors. *New J. Chem.* **2016**, *40*, 9187–9193. [CrossRef]
24. Gao, J.; Cong, J.; Wu, Y.; Sun, L.; Yao, J.; Chen, B. Bimetallic hofmann-type metal-organic framework nanoparticles for efficient electrocatalysis of oxygen evolution reaction. *ACS Appl. Energy Mater.* **2018**, *1*, 5140–5144. [CrossRef]
25. Vuong, G.T.; Pham, M.H.; Do, T.O. Direct synthesis and mechanism of the formation of mixed metal Fe 2Ni-MIL-88B. *CrystEngComm* **2013**, *15*, 9694–9703. [CrossRef]
26. Vuong, G.T.; Pham, M.H.; Do, T.O. Synthesis and engineering porosity of a mixed metal Fe2Ni MIL-88B metal-organic framework. *Dalton Trans.* **2013**, *42*, 550–557. [CrossRef]
27. Trinh, N.D.; Hong, S.-S. Photocatalytic decomposition of methylene blue over MIL-53(Fe) prepared using microwave-assisted process under visible light irradiation. *J. Nanosci. Nanotechnol.* **2014**, *15*, 5450–5454. [CrossRef]
28. Nguyen, V.; Nguyen, T.; Bach, L.; Hoang, T.; Bui, Q.; Tran, L.; Nguyen, C.; Vo, D.-V.; Do, S. Effective photocatalytic activity of mixed Ni/Fe-base metal-organic framework under a compact fluorescent daylight lamp. *Catalysts* **2018**, *8*, 487. [CrossRef]
29. Schejn, A.; Falk, V.; Mozet, K.; Schneider, R.; Aboulaich, A.; Balan, L.; Lalevée, J.; Medjahdi, G.; Aranda, L. Cu2+-doped zeolitic imidazolate frameworks (ZIF-8): Efficient and stable catalysts for cycloadditions and condensation reactions. *Catal. Sci. Technol.* **2015**, *5*, 1829–1839. [CrossRef]
30. Wu, M.S.; Chen, F.Y.; Lai, Y.H.; Sie, Y.J. Electrocatalytic oxidation of urea in alkaline solution using nickel/nickel oxide nanoparticles derived from nickel-organic framework. *Electrochim. Acta* **2017**, *258*, 167–174. [CrossRef]
31. Wu, M.S.; Chen, C.Y.; Chen, Y.R.; Shih, H.C. Synthesis of bimodal mesoporous carbon with embedded nickel nanoparticles through pyrolysis of nickel-organic framework as a counter-electrode catalyst for dye-sensitized solar cells. *Electrochim. Acta* **2016**, *215*, 50–56. [CrossRef]
32. Nguyen, V.H.; Bach, L.G.; Do, S.T.; Thuong, N.T.; Nguyen, T.D. Photoluminescence properties of Eu-doped MIL-53(Fe) obtained by solvothermal synthesis. *J. Nanosci. Nanotechnol.* **2018**, *19*, 1148–1150. [CrossRef] [PubMed]
33. Haque, E.; Khan, N.A.; Park, H.J.; Jhung, S.H. Synthesis of a metal-organic framework material, iron terephthalate, by ultrasound, microwave, and conventional electric heating: A kinetic study. *Chem. A Eur. J.* **2010**, *16*, 1046–1052. [CrossRef] [PubMed]
34. Sun, Q.; Liu, M.; Li, K.; Han, Y.; Zuo, Y.; Chai, F.; Song, C.; Zhang, G.; Guo, X. Synthesis of Fe/M (M = Mn, Co, Ni) bimetallic metal organic frameworks and their catalytic activity for phenol degradation under mild conditions. *Inorg. Chem. Front.* **2017**, *4*, 144–153. [CrossRef]
35. Vu, T.A.; Le, G.H.; Dao, C.D.; Dang, L.Q.; Nguyen, K.T.; Nguyen, Q.K.; Dang, P.T.; Tran, H.T.K.; Duong, Q.T.; Nguyen, T.V.; et al. Arsenic removal from aqueous solutions by adsorption using novel MIL-53(Fe) as a highly efficient adsorbent. *RSC Adv.* **2015**, *5*, 5261–5268. [CrossRef]
36. Feng, X.; Chen, H.; Jiang, F. In-situ ethylenediamine-assisted synthesis of a magnetic iron-based metal-organic framework MIL-53(Fe) for visible light photocatalysis. *J. Colloid Interface Sci.* **2017**, *494*, 32–37. [CrossRef]

 © 2019 by the authors. Licensee MDPI, Basel, Switzerland. This article is an open access article distributed under the terms and conditions of the Creative Commons Attribution (CC BY) license (http://creativecommons.org/licenses/by/4.0/).

Article

An Experimental Approach on Industrial Pd-Ag Supported α-Al$_2$O$_3$ Catalyst Used in Acetylene Hydrogenation Process: Mechanism, Kinetic and Catalyst Decay

Ourmazd Dehghani, Mohammad Reza Rahimpour * and Alireza Shariati

Department of Chemical Engineering, Shiraz University, Shiraz 71345, Iran; ourmazd1@yahoo.com (O.D.); shariati@shirazu.ac.ir (A.S.)
* Correspondence: rahimpor@shirazu.ac.ir; Tel.: +98-713-2303071

Received: 4 February 2019; Accepted: 19 February 2019; Published: 5 March 2019

Abstract: The current research presents an experimental approach on the mechanism, kinetic and decay of industrial Pd-Ag supported α-Al$_2$O$_3$ catalyst used in the acetylene hydrogenation process. In the first step, the fresh and deactivated hydrogenation catalysts are characterized by XRD, BET (Brunauer–Emmett–Teller), SEM, TEM, and DTG analyses. The XRD results show that the dispersed palladium particles on the support surface experience an agglomeration during the reaction run time and mean particle size approaches from 6.2 nm to 11.5 nm. In the second step, the performance of Pd-Ag supported α-Al$_2$O$_3$ catalyst is investigated in a differential reactor in a wide range of hydrogen to acetylene ratio, temperature, gas hourly space velocity and pressure. The full factorial design method is used to determine the experiments. Based on the experimental results ethylene, ethane, butene, and 1,3-butadiene are produced through the acetylene hydrogenation. In the third step, a detailed reaction network is proposed based on the measured compounds in the product and the corresponding kinetic model is developed, based on the Langmuir-Hinshelwood-Hougen-Watson approach. The coefficients of the proposed kinetic model are calculated based on experimental data. Finally, based on the developed kinetic model and plant data, a decay model is proposed to predict catalyst activity and the parameters of the activity model are calculated. The results show that the coke build-up and condensation of heavy compounds on the surface cause catalyst deactivation at low temperature.

Keywords: acetylene hydrogenation; kinetic model; catalyst decay; process modeling

---

1. Introduction

Generally, ethylene is one of the most important building blocks in the chemical industry, which is widely used to produce a wide range of products and intermediates, such as polyethylene, ethylene oxide, ethylbenzene, and ethylene dichloride [1,2]. Although the catalytic conversion of hydrocarbons to ethylene is beneficial, the steam thermal cracking of ethane, LPG, naphtha, and gasoline is the most popular method to produce ethylene. Typically, a wide range of hydrocarbons is produced in the thermal cracking process. Acetylene as a by-product of cracking unit has an enormous effect on the quality of product and must be removed from the olefin streams prior to further processing [3]. Typically, the minimum required purity of ethylene in the polymerization processes to produce polyethylene is about 99.90% and the maximum allowable limit of acetylene is 5 ppm known as polymer-grade ethylene. Acetylene decreases the catalyst activity in the ethylene polymerization unit, and can produce metal acetylides as explosive compartments. In this regard, several technologies have been proposed to decrease the acetylene concentration in the effluent product from thermal cracking furnaces, including acetylene hydrogenation to ethylene and acetylene separation from the main stream [4]. Since the separation process is expensive and dangerous, the catalytic hydrogenation is more popular and attractive.

1.1. Hydrogenation Catalysts

Catalyst selection and preparation is one of the most important stages in process design and development. Generally, Pd, Pd-Ag, and Pd-Au supported on α-Al_2O_3 have been designed to use in the industrial acetylene hydrogenation process [5–7]. Ravanchi et al. reviewed the theoretical and practical aspects of catalysis for the selective hydrogenation of acetylene to ethylene and the potential ways to improve catalyst formulation [8]. Bos et al. investigated the kinetics of the acetylene hydrogenation on a commercial Pd catalyst in a Berty type reactor [9]. The considered reaction network consists of acetylene hydrogenation and ethylene hydrogenation reactions. They proposed different rate expressions and calculated the parameters of rates, based on the experimental data. The results showed that the classical Langmuir-Hinshelwood rate expressions could not fit the data well, when there is a small amount of carbon monoxide in the feed stream. Borodziński focused on the hydrogenation of acetylene and mixture of acetylene and ethylene on the palladium catalyst [10]. The results showed that two different active sites are detectable based on the palladium size. The results showed that, although acetylene and hydrogen are adsorbed on the small active site, ethylene did not adsorb, due to steric hindrance. In addition, all reactants were adsorbed on the large sites and butadiene as coke precursor was produced on that site. Zhang et al. investigated the performance of Pd-Al_2O_3 nano-catalyst in the acetylene hydrogenation [11]. The results showed that dispersing Ag as a promoter on the catalyst surface increases ethylene selectivity from 41% to 60% at 100 °C. Typically, adding Au to Pd-Al_2O_3 can tolerate carbon monoxide concentration swing, and improve the selectivity, and temperature resistant [12]. Schbib et al. investigated the kinetics of acetylene hydrogenation over Pd-Al_2O_3 in the presence of a large excess of ethylene in a laboratory flow reactor [13]. They claimed that C_2H_2 and C_2H_4 compounds are adsorbed on the same site and they react with the adsorbed hydrogen atoms to form C_2H_4, and C_2H_6, respectively. It appeared that the presence of a trace amount of silver on Pd-Al_2O_3 catalyst decreases the rate of ethylene hydrogenation as a side reaction [14]. Khan et al. studied adsorption and co-adsorption of ethylene, acetylene, and hydrogen on Pd-Ag, supported on α-Al_2O_3 catalyst by temperature programmed desorption [15]. The TPD (temperature programmed desorption) results showed that, although the presence of Ag on the catalyst suppresses overall hydrogenation activity, it increased the selectivity towards ethylene [16]. Pachulski et al. investigated the effect green oil formation and coke build-up has on the deactivation of Pd-Ag, supported on α-Al_2O_3 catalyst, applied in the C_2-tail end-selective hydrogenation [17]. It was found that the catalyst contains low Ag to Pd ratio presents the highest long-term stability. The characterization results showed that the regenerated samples present the same stability. Currently, the use of non–toxic and inexpensive metals such as Fe, Ti, Cu or Zr, instead of Pd and Ag based commercial catalysts is an attractive topic. In this regard, Serrano et al, focused on the embedding FeIII on an MOF to prepare an efficient catalyst for the hydrogenation of acetylene under front–end conditions [18]. The experimental results showed that the prepared catalyst presents similar activity to Pd catalyst and could control acetylene concentration at the desired level.

1.2. Hydrogenation Method

The Front-End and Tailed-End are two common methods in acetylene hydrogenation, which differ in the reactor structure and process arrangements. In the Front-End method, the feed stream, which may contain up to 40% hydrogen, directly enters into the hydrogenation reactor and feeds temperature, is the only manipulated variable. Gobbo et al. modeled and optimized the Front-End acetylene hydrogenation process considering catalyst deactivation [19]. They calculated the dynamic optimal trajectory of feed temperature to control acetylene concentration at desired level. In the Tail-End method, hydrogen is separated from the effluent stream from steam cracker. In this method, the feed temperature and hydrogen concentration in the feed stream are manipulated variables. Aeowjaroenlap et al. modeled the Tailed-End hydrogenation reactors, based on the mass and energy balance equations at dynamic condition [20]. To obtain the optimum operating condition, a single objective dynamic optimization problem was formulated to maximize process economics. The inlet temperature and hydrogen concentration were selected as the decision variables. The results showed that applying optimal operation condition on the system increases process economics about 10%.

1.3. Reactor Arrangement

Typically, the acetylene hydrogenation process contains four catalytic beds, namely Lead and Guard Beds. The philosophy of guard bed is the sensitivity of downstream units to acetylene and the decreasing acetylene concentration to the desired level [21]. The coke build-up on the catalyst surface decreases activity and increases acetylene concentration in outlet stream from Guard bed gradually. In this regard, two beds are in operation, while two other beds are in standby or regeneration modes. Dehghani et al. modified the reaction-regeneration cycles and the reactor arrangement in the acetylene hydrogenation process to decrease energy consumption, and improve catalyst lifetime [22]. The feasibility of the proposed configuration was proved based on a theoretical framework.

1.4. Research Outlook

In this research, the reaction mechanism and kinetics of acetylene hydrogenation over the industrial Pd-Ag supported on α-Al_2O_3 is investigated in a lab-scale packed bed reactor, considering GHSV (gas hourly space velocity), hydrogen to acetylene ratio, pressure, and temperature as independent variables. The full factorial design of experiment method based on the cubic pattern is used to determine the number and condition of experiments. The fresh and deactivated catalysts are characterized by XRD, BET, SEM, TEM and DTG analyses. In addition, a detail reaction network is proposed and correspond kinetic model is developed based on the Langmuir-Hinshelwood- Hougen-Watson approach. Then, the Tail-End hydrogenation reactors in Jam Petrochemical Complex are modeled based on the mass and energy balance equations at dynamic conditions. Based on the developed model and available plant data, a decay model is proposed to predict catalyst activity. Then, the accuracy of the model is proved at steady and dynamic conditions.

2. Experimental Method

2.1. Catalyst Preparation

In this research, the performance of industrial Pd-Ag supported α-Al_2O_3 catalyst is investigated in a lab-scale reactor. The catalyst OleMax® 201 manufactured by SÜD-CHEMIE (Germany, Munich) is supplied from Jam Petrochemical Complex in Iran. It is a high performance, stable, and flexible catalyst to maximize olefins production through acetylene hydrogenation. Table 1 shows the specification of fresh catalyst.

Table 1. The specification of fresh catalyst.

Bulk Density (kg m^{-3})	720
Size (mm)	2–4
Shape	Sphere
Pd content (ppm)	300
Ag to Pd ratio	6
Particle porosity (%)	60–70%
Particle tortuosity	2.5
BET Surface Area (m^2 g^{-1})	30.1062
BJH Adsorption average pore diameter (Å)	291.218
Thermal conductivity of catalyst (W m^{-1} k^{-1})	0.29

The industrial catalyst is prepared by impregnation method and Pd and Ag are dispersed on the catalyst separately. Before tests, the catalyst is activated by removing water from the pores and subsequent reduction of palladium oxide on the support to palladium black. The removal of water is carried out by purging nitrogen through the reactor at 150 °C for 2 h. The reduction is conducted by hydrogen-diluted stream at 150 °C. After reduction, the catalyst is cooled to ambient temperature by the nitrogen purging.

2.2. Catalyst Characterization

The supplied catalyst is characterized by BET, TGA, XRD, SEM, and TEM analysis. BET analysis is used to measure the specific surface area and the pore size distribution of catalyst. The SEM test is used to analysis the surface and morphology of catalyst by scanning the surface with a focused beam of electron. TGA is a thermal method used to investigate the stability of a catalyst during heating. In this regard, the mass of the catalyst is measured over time during the heating. The XRD technique is an analytical tool used to determine the phase and dimension of crystalline material. In the present research, the SEM and TEM analyses were performed by using Philips XL 30 (FEI Company, Hillsboro, OR, USA) and FEI Tecnai G^2 F20 (FEI Company, Hillsboro, OR, USA), respectively. The XRD pattern of the catalyst was recorded on a Rigaku D/Max-2500 (Rigaku, Austin, TX, USA) diffractometer at a scanning speed of 4 min^{-1} over the 2θ range of 10–80°. The TGA and DTA (differential thermal analysis) analysis of the fresh and deactivated catalysts were performed by Mettler Toledo Model 2007. The nitrogen adsorption and desorption tests were measured by Quanta chrome Autosorb at 70 K. The specific surface area of the catalyst was calculated by the Brunauer–Emmett–Teller equation. In addition, the Horvath and Kawazoe equation were used to calculate the pore size and volume of catalyst particles.

The supplied feed stream contains acetylene, ethylene, and ethane contaminated with a trace of propylene and methane. After regulation of the temperature and flow rate, feed stream enters to the reactor and passes over the ceramic ball and catalyst layers. The ceramic ball layer is considered to uniform distribution of feed along the catalytic bed. To detect the product distribution, the effluent is attached to the gas chromatography and product composition is measured on-line. Figure 1 shows the designed reactor to investigate kinetic of acetylene hydrogenation.

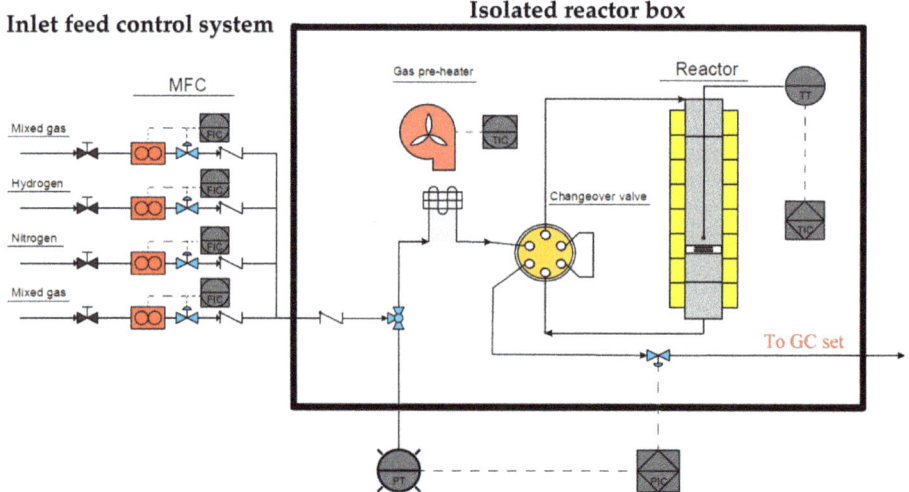

Figure 1. The designed reactor to investigate kinetic of acetylene hydrogenation.

2.3. Experimental Apparatus

The designed reactor is a stainless steel cylindrical chamber with the inner diameter of 9 mm and the length of 300 mm. To control the reactor temperature, feed temperature, flow rate, and pressure, the setup was equipped by a jacket heater, heating blower, MFC model F-231M, made by Bronkhoest, pressure sensor model DP2-21 and backpressure regulator 1315G2Y, made by Hoke (prentice HALL, Upper Saddle River, NJ, USA), respectively. In the designed tests, the feed stream was supplied from industrial acetylene hydrogenation unit in Jam Petrochemical Complex. Table 2 shows the composition of the feed stream in the acetylene hydrogenation unit of Jam Petrochemical Complex.

Table 2. The composition of the feed stream.

Methane	0.014
Acetylene	0.738
Ethylene	64.594
Propane	0.002
Propylene	0.199
Ethane	34.449
Other C_{4s}	0.0023
MAPD	0.0005
Cyclopropane	<0.0001
C_{5+} Hydrocarbons	<0.0001
1,3 Butadiene	<0.0001

3. Kinetic Modeling

3.1. Experiment Design

During the past decade, several designs of experimental approaches have been developed to reduce the numbers of experiments [23]. In the current research, the factorial design method, based on the cubic pattern, is used to determine the experiments. In statistics, the full factorial is an experimental design method, whose design consists of two or more factors, each with discrete possible levels. In the first step, the effective parameters, ranges, and levels are selected to cover a wide range of operating condition. The considered independent variables are temperature, pressure, hydrogen to acetylene ratio, and GHSV. The fraction of products in the outlet stream is selected as the objective function. Table 3 shows the variation range and the number of data points. Considering full factorial design method, 216 independent experiments are designed.

Table 3. The variation range and number of data points.

	Lower	Upper	Number of Levels
Hydrogen to acetylene ratio	0.5	1.5	3
Pressure (Bar)	15	20	3
Temperature	35	60	4
GHSV	2600	6200	6

3.2. Reaction Mechanism

In this research, acetylene conversion to ethylene, ethane, butenes, and butadiene are considered as independent reactions in the considered network. Typically, the ethylene and acetylene could be adsorbed on the catalyst surface as:

$$C_2H_2 \xleftrightarrow{\text{Adsorbed as}} \begin{cases} -CH=CH- & di-\sigma-Adsorbed \\ CH\equiv CH & \pi-Complex \\ -CH-CH_3 & Ehthylidene \\ -CH=CH_2 & Vinyl \\ -C-CH_3 & Ethylidene \\ =C=CH_2 & Vinylidene \end{cases} \quad (1)$$

$$C_2H_4 \xleftrightarrow{\text{Adsorbed as}} \begin{cases} -CH_2-CH_2- & di-\sigma-Adsorbed \\ CH_2=CH_2 & \pi-Complex \\ -CH-CH_3 & Ehthylidene \\ -CH=CH_2 & Vinyl \\ -C-CH_3 & Ethylidene \end{cases} \quad (2)$$

Based on the density functional theory, selective acetylene hydrogenation to ethylene considering vinyl layer as the intermediate is the most dominant mechanism [24]. Based on the considered reaction mechanism, hydrogen is adsorbed on the catalyst surface as:

$$H_2 + 2S \leftrightarrow 2H-S \quad (3)$$

In addition, acetylene is adsorbed on the surface and reacts with adsorbed hydrogen to produce ethylene:

$$C_2H_2(g) + S \underset{-H}{\overset{+H}{\rightleftarrows}} CH_2\,CH-S \xrightarrow{+H} CH_2CH_2(g) + s \quad (4)$$

In addition, ethylene in the gas phase is adsorbed on the surface and reacts with adsorbed hydrogen in two steps to produce ethane as:

$$H_2CH_2\,(g) + S \longleftrightarrow CH_2CH_2-S + H-S \longleftrightarrow CH_3CH_2-S+S \quad (5)$$

$$CH_3CH_2-S+H-S \longleftrightarrow CH_3CH_3(g) + 2S \quad (6)$$

In general, there are two possible pathways to produce butadiene. According to the first path:

$$CHCH(g) + S \longleftrightarrow CH_2\,CH-S \underset{-H}{\overset{+H}{\rightleftarrows}} CH_3\,CH-S \quad (7)$$

$$HCH(g) + S \longleftrightarrow CH_2\,C-S \quad (8)$$

$$CH_3\,CH-S + CH_2\,C-S \longrightarrow CH_2CH\,CHCH_2(g) \quad (9)$$

According to the second path:

$$C_2H_2(g) + S \underset{-H}{\overset{+H}{\rightleftarrows}} CH_2\,CH-S \quad (10)$$

$$C_2H_2(g) + S \underset{-H}{\overset{+H}{\rightleftarrows}} \check{C}H_2\,CH-S \quad (11)$$

$$CH_2\,CH-S + \check{C}H_2\,CH-S \longrightarrow CH_2CHCHCH_2(g) + 2\,S \quad (12)$$

Typically, 1,3-butadiene could be found in two different states, including in the gas phase and on the solid surface. In the first state, butadiene is detected in the outlet stream from the reactor, while the second state is a complex state that causes oligomer production. The produced oligomer is precipitated on the catalyst surface and leads to deactivation of the catalyst by blocking active sites [17,25]. Thus, to investigate butadiene and oligomer formation, the outlet gas stream from the reactor is analyzed by

GC-mass and PIONA. The results of GC-mass has been presented in the Supplementary Materials (Data Set 3). The mechanism of 1-butene formation on the catalyst surface is:

$$C_2H_2(g) + S \underset{-H}{\overset{+H}{\rightleftarrows}} CH_2CH-S \underset{-H}{\overset{+H}{\rightleftarrows}} CH_3CH-S + CH_2CH-S \longrightarrow CH_2CHCH_2CH_3(g) \quad (13)$$

In addition, the mechanism of cis-2-butane and trans-2-butane formation is as:

$$C_2H_2(g) + S \underset{-H}{\overset{+H}{\rightleftarrows}} CH_2CH-S \underset{-H}{\overset{+H}{\rightleftarrows}} H_3CH-S + CH_3CH-S \longrightarrow CH_3CHCHCH_3(g) \quad (14)$$

3.3. Kinetic Model

In this section, based on the considered mechanism, a reaction network comprising six reactions is selected. The considered reactions are as follows:

$$C_2H_2 + H_2 \rightarrow C_2H_4 \quad (15)$$

$$C_2H_4 + H_2 \rightarrow C_2H_6 \quad (16)$$

$$2\,C_2H_2 + H_2 \rightarrow C_4H_6 \quad (17)$$

$$2\,C_2H_2 + 2H_2 \rightarrow Cis-C_4H_8 \quad (18)$$

$$2\,C_2H_2 + 2H_2 \rightarrow Trans-C_4H_8 \quad (19)$$

$$2\,C_2H_2 + 2H_2 \rightarrow 2-C_4H_8 \quad (20)$$

To simplify the acetylene hydrogenation to 1-butene, cis-2-butene and trans-2-butene reactions are lumped to acetylene hydrogenation to butene group. Based on the considered reaction network and Langmuir-Hinshelwood-Hougen-Watson mechanism, a detail kinetic model is proposed to predict the rate of reactions and the coefficients of the considered model are calculated based on experimental data [26]. The considered rate of reactions is as follows:

$$r_i = \frac{k_i \prod P_j^n}{(1 + \sum K_j P_j)^m} \quad (21)$$

3.4. Deactivation Model

In general, the five intrinsic mechanisms of catalyst decay are poisoning, fouling, thermal degradation, chemical degradation, and mechanical failure [27]. Poisoning and thermal degradation are generally slow and irreversible, while fouling by coke and carbon is rapid and reversible. Generally, one of the main challenges in the acetylene hydrogenation process is catalyst deactivation, by coking and increasing acetylene concentration in the product stream. Typically butene and butadiene, as side products in acetylene hydrogenation, has led to oligomer and green oil formation on the catalyst surface [28]. The adsorbed acetylene and produced 1,3-butadiene react on the catalyst surface and green oil is produced. The deposited oligomers and green oil on the catalyst gradually reduce the catalyst activation during the process run-time [29]. The proposed correlations in the literature, that predict catalyst activity lack accuracy, so applying these activity equations in the model results in a notable error between simulation results and plant data. In this research, a power low decay model, modified by feed concentration, to account for coke formation, is proposed to calculate the catalyst activity. The considered deactivation model is as follows:

$$\frac{da}{dt} = k_d e^{-(\frac{E_d}{RT})} \times a^n \times C^m \quad (22)$$

The proposed decay model is applied in the dynamic model and the available plant data are used to calculate the activity parameters, considering the absolute difference between plant data and simulation results as the objective function.

4. Process Modeling

In this section, the industrial two-stage acetylene hydrogenation processes are modeled on the mass and energy balance equation at pseudo-steady state conditions. The adopted assumptions in the considered model are:

- Pseudo-steady state condition;
- the plug flow pattern in the reactor;
- negligible concentration and temperature gradients in the catalyst particle;
- negligible radial mass and energy diffusion;
- negligible mass and heat transfer in the longitudinal direction; and
- adiabatic conditions.

The gas is at non-ideal condition and Redlich-Kwong equation of state is considered to predict gas phase property due to high pressure and low temperature conditions. The mass, energy and moment balance equations in the bed could be explained as follows:

$$\frac{dn_A}{dz} = a \sum_{i}^{N} v_i r_i \rho_B A \tag{23}$$

$$\frac{dT}{dz} = \frac{A\rho_B}{n_t C_p} \sum_{i}^{M} r_j \times (-\Delta H_j) \tag{24}$$

$$\frac{dP}{dz} = \frac{150 \mu V (1-\varepsilon)^2}{\varphi^2 D_p^2 \varepsilon^3} + \frac{1.75 \rho V^2 (1-\varepsilon)}{\varphi D_p \varepsilon^3} \tag{25}$$

Combining balance equations, kinetic model, auxiliary equations to predict physical and chemical properties, and activity models result in a set of algebraic and partial differential equations. In the developed model, the mass and energy balance equations are written at a steady state condition, while the activity equation is a dynamic model.

5. Optimization Problem

In this research, to calculate the coefficients of the proposed kinetic and activity models, an optimization problem was formulated to minimize the absolute difference of model results with experimental data. The Genetic Algorithm is a powerful method in global optimization and has been selected to handle formulated optimization problems and obtain the coefficients of kinetic and activity models. Genetic algorithms are the most popular evolutionary algorithm, that is inspired by natural selection of the fittest populations to reproduce and move to the next generation [30]. In each generation the fittest population are attained by three operators, consist of selection, crossover and mutation. In the kinetic section, the reaction rate is calculated numerically and the absolute difference between calculated reaction rate by the model and measured rates are minimized. The considered objective function is as:

$$AMRE = \frac{1}{N} \sum_{i=1}^{i=N} \frac{|y_{exp}(i) - y_{model}(i)|}{y_{exp}(i)} \times 100 \tag{26}$$

To calculate the coefficients of the considered activity model, the outlet acetylene concentration from guard and lead beds is measured and compared with the calculated acetylene concentration by the model. The considered data consists of 48 data point during the process run time.

6. Results and Discussions

6.1. Catalyst Characteristics

In this section, the results of SEM, TEM, DTG-TGA, XRD, and BET of fresh and spent catalysts, are presented. It is mentioned that a used catalyst in the plant is named spent catalyst. As mentioned, to investigate the surface morphology, the BET analysis is performed on the fresh and spent catalysts. Table 4 shows the BET results of fresh and spent catalysts. The obtained results reveal that BET surface area of fresh and spent catalysts are 24.75, and 30.11 $m^2 g^{-1}$, respectively. In addition, the mean pore diameter of fresh and deactivated catalysts are 235.5, and 191.2 Å, respectively. It concludes that, from BET analysis, there is coke build-up on the internal pores and pore blockage by coke decrease mean pore diameter. In addition, coke build-up on the external surface of the catalyst and increases surface area. Figure 2 shows the results of nitrogen adsorption and desorption on the fresh catalyst.

Table 4. The BET results of fresh and spent catalysts.

	Fresh	Spent
BET surface area ($m^2 g^{-1}$)	24.75	30.11
Langmuir surface area ($m^2 g^{-1}$)	34.13	41.86
External surface area ($m^2 g^{-1}$)	20.15	27.27
Micro pore area ($m^2 g^{-1}$)	4.6018	2.84
Adsorption average pore width (Å)	235.52	191.17
Adsorption cumulative volume of pores ($cm^3 g^{-1}$)	0.229	0.218

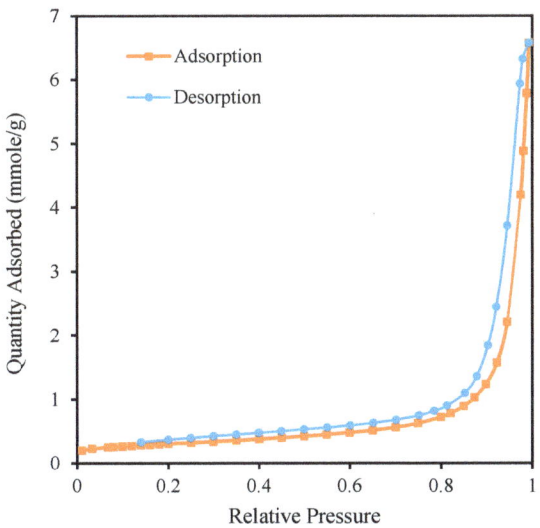

Figure 2. The BET results of the fresh catalyst.

It appears that increasing pressure increases the nitrogen adsorption on the catalyst surface and adsorption pattern, in accordance with the Isotherm Type III. This could be applied on systems in which the interaction between adsorbate molecules is stronger than that between adsorbate and adsorbent. Based on the Isotherm Type III, the uptake of gas molecules is initially slow, and until surface coverage is sufficient, so that the interactions between adsorbed and free molecules begins to dominate the process.

Typically, the XRD analysis is used to identify the crystalline morphology and dimensions of support. Figure 3 shows the XRD results of the fresh catalyst. The broad peak means poor crystalline morphology and the sharp ones indicate a well-crystallized sample. Based on the XRD analysis, the peak is at 32.75°, which proves the presence of Ag_2O particles on the support surface, while peaks at 36.7°, 63.98°, and 67.46° show Ag conversion to AgO. In addition, the peaks at 38.9° and 66.2° show the dispersion of Pd on the catalyst surface. Based on these XRD results, the Al_2O_3 mean crystal size is 24.5 nm.

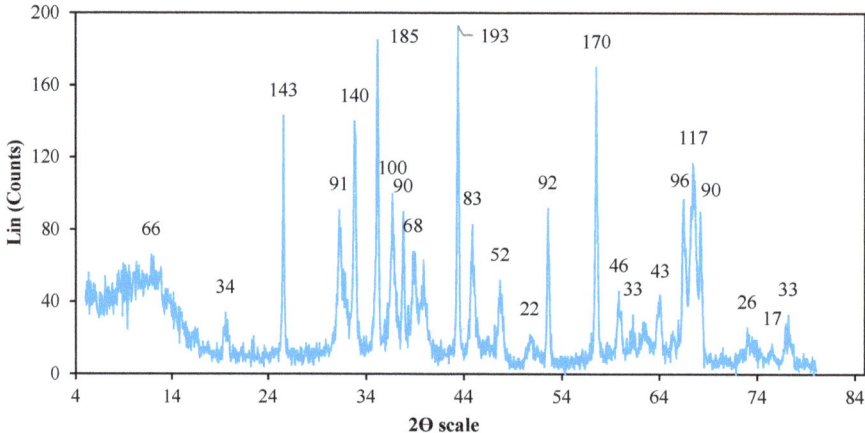

Figure 3. The XRD results of the fresh catalyst.

Figure 4 shows the SEM images of fresh and spent catalysts. The results of SEM images reveal that the bright trace of palladium metal, in fresh catalyst, changes in the dark in the spent catalyst. The darkening of catalyst proves the formation of polymeric compounds and coke build-up on the catalyst surface, which reduces the activity of the catalysts especially. Indeed the surface of the fresh catalyst is completely covered by coke. It concludes from SEM and BET tests that coke build-up on the external surface of the catalyst increases the surface area.

Figure 4. SEM images of fresh catalysts and spent catalysts. (**a**) Fresh catalyst, (**b**) Spent catalyst.

In addition, Figure 5a,b presents TEM image and particle size distribution of fresh and spent catalysts. The results show that palladium particles experience an agglomeration during the reaction run time and mean particle size (c), (d) approaches from 6.2 nm to 11.5 nm. Increasing size of palladium particles, during the run-time, reduces the active sites and results in lower catalyst activity.

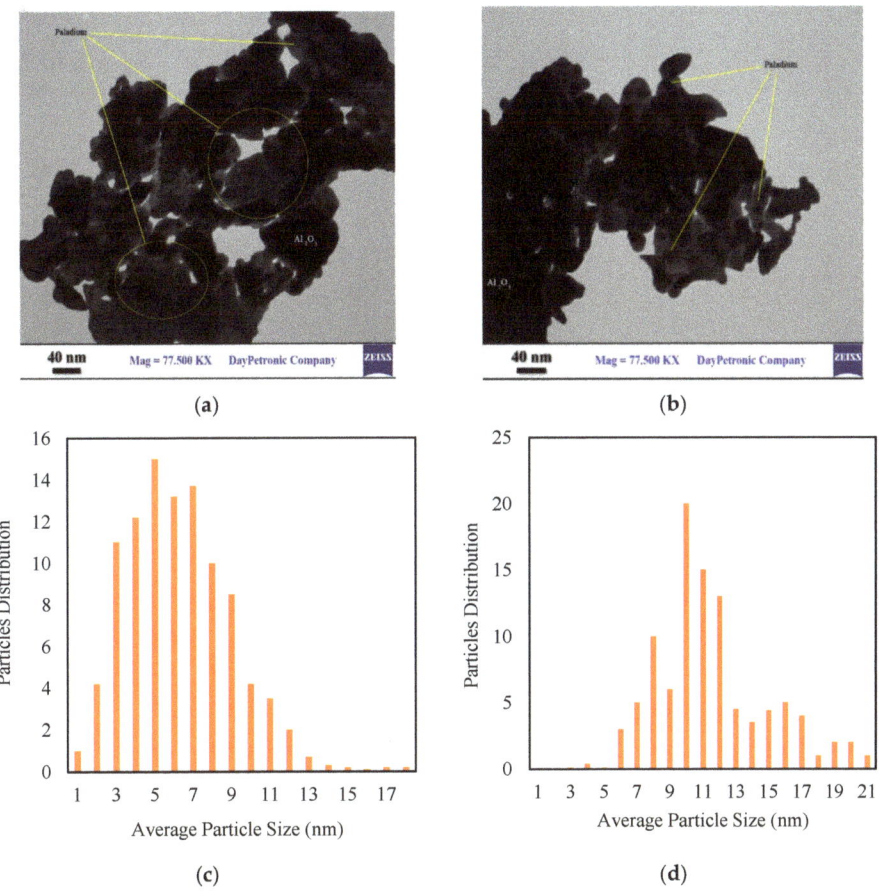

Figure 5. (a,b) TEM images of fresh and spent catalysts, (c,d), particle size distribution of fresh and spent catalysts.

Figure 6a,b shows the TGA and DTG results of fresh, spent and regenerated catalysts. Generally, the fresh and regenerated catalysts do not experience weight loss during the TGA test. However, the oxidation of Pd and Ag atoms to PdO and AgO, increases catalyst weight by about 0.6%. The TGA results of deactivated catalysts shows that, increasing the temperature up to 500 °C decreases sample weight gradually and after that, catalysts do not experience weight loss. Typically, coke burning during the TGA analysis is the main reason for the decreased catalyst weight. In addition, it is concluded that the coke is completely burned through catalyst heating up to 500 °C. The two minimum points at 310 and 515 °C on the DTG curve of spent catalyst proves the presence of two different coke types on the catalyst surface. The produced amorphous coke on the external surface of catalyst burns in temperature range of 300 to 400 °C, while the crystalline coke and produced coke in the pores burn in range of 450 to 650 °C.

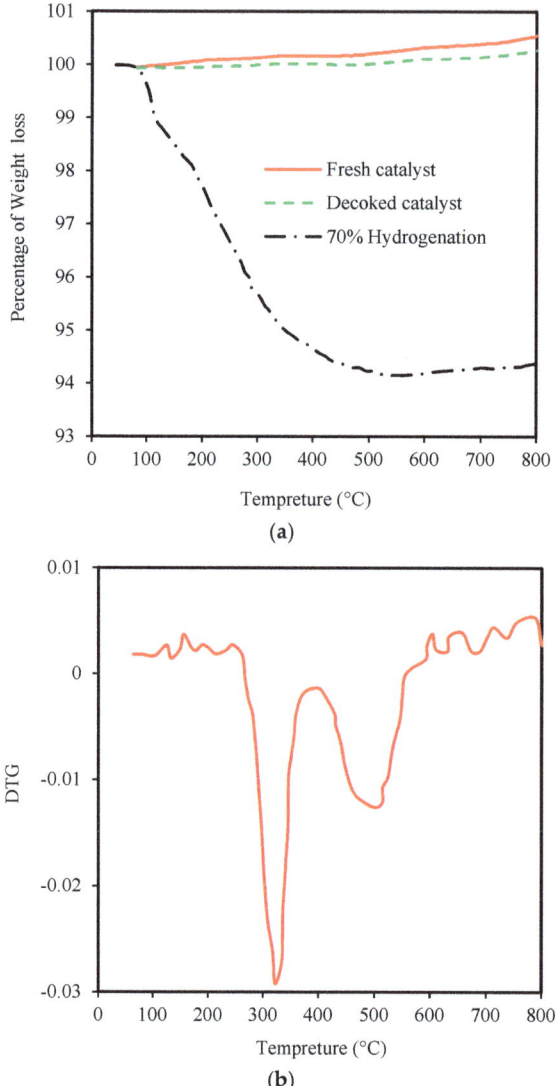

Figure 6. (**a**) TGA results of different catalysts and (**b**) DTG results of coked catalysts.

6.2. Results of Kinetic Model

As previously mentioned, 216 experiments have been designed to find the effect of parameters on the acetylene conversion and product distribution. The list of experiments and results have been tabulated in Supplementary Data Set 1. In this section, the effect of GHSV, temperature, pressure, and hydrogen to acetylene ratio on acetylene conversion, ethylene selectivity, and product distribution is presented.

6.2.1. Effect of GHSV

Figure 7a,b shows the effect of gas hourly space velocity on acetylene conversion, ethylene selectivity, and product distribution. The GHSV is the ratio of gas flow rate in standard condition to the volume of catalyst in the bed. Although increasing GHSV reduces residence time in the reactor, it decreases mass transfer resistance in the bed. The experiments show that increasing GHSV results in higher ethylene selectivity and lower acetylene conversion. Although butene group and 1,3-butadiene could be detected in the outlet stream from the reactor, 1-butene is the dominant side product. It appears that GHSV has a considerable effect on the 1-butene formation and increasing GHSV from 2500 to 6200 decreases 1-butene mole fraction from 0.02 to 0.002. It is concluded that increasing the GHSV led to a reduction in residence time and consequently enhances the risk of leaving unreacted acetylene from the reactor.

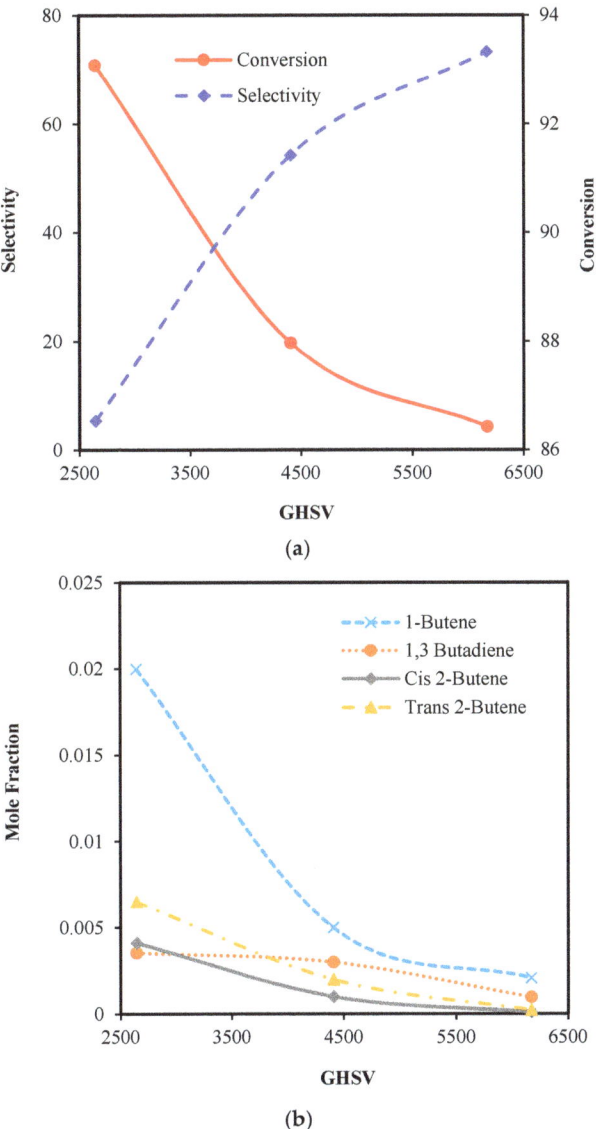

Figure 7. Effect of GHSV on (**a**) acetylene conversion and ethylene selectivity, and (**b**) product distribution at 15 bar, 35 °C and hydrogen to acetylene ratio 0.5.

6.2.2. Effect of Pressure

Typically, pressure is one of the most effective parameters influencing acetylene conversion and ethylene selectivity. In detail, the pressure could change, both the adsorption coefficients of catalysts and the partial pressure of the participated components on the catalyst. Figure 8a,b shows the effect of operating pressure on acetylene conversion, ethylene selectivity, and product distribution. Based on the experiments, although increasing pressure improves acetylene conversion, selectivity decreases sharply in pressure range 15–18 bar and after that decreases gradually. Typically, pressure increases the diffusivity and adsorption of components in the surface of the catalyst and results in a higher conversion factor. It appears that the main side product is 1-butane and increasing operating pressure increases the rate of side products in the system.

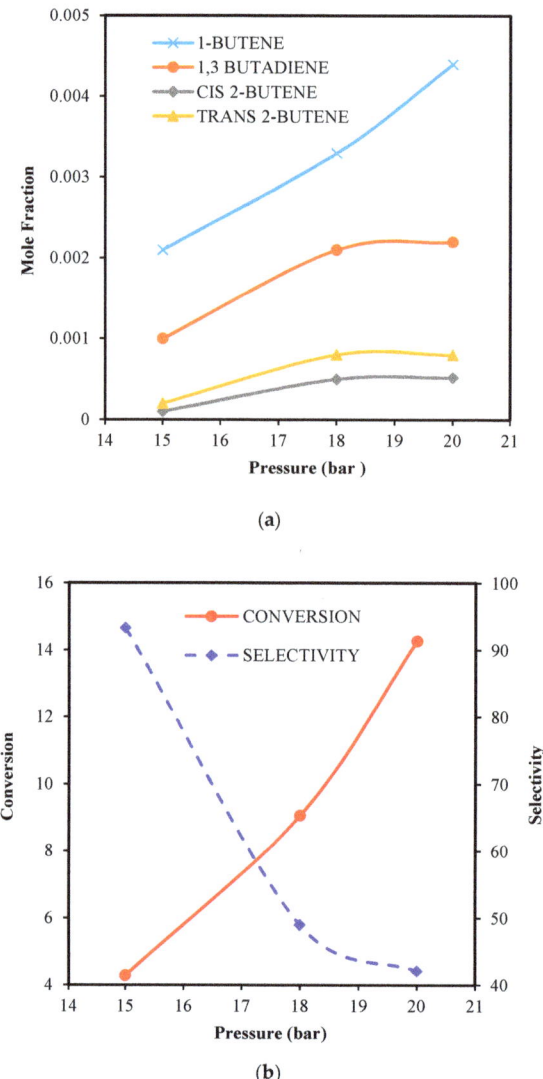

Figure 8. Effect of pressure on (**a**) acetylene conversion, ethylene selectivity, and (**b**) product distribution at 35 °C and hydrogen to acetylene ratio 0.9 and GHSV 3500 1/h.

6.2.3. Effect of Temperature

Typically, temperature is the most important parameter and has a direct effect on the selectivity and conversion. Increasing the temperature improves the reaction rate and shifts the reversible exothermic reactions toward lower equilibrium conversion. Figure 9a,b shows the effect of operating temperature on acetylene conversion, ethylene selectivity, and product distribution. Although applying higher temperature increases the rate of acetylene conversion as the main reaction, it increases rate of side reactions. Typically, applying high temperatures on the system has a considerable effect on the side reactions, so butene and butadiene formation increase sharply. Since the increasing temperature decreases selectivity, the effect of operating temperature on side reactions is more significant compared to the main reaction.

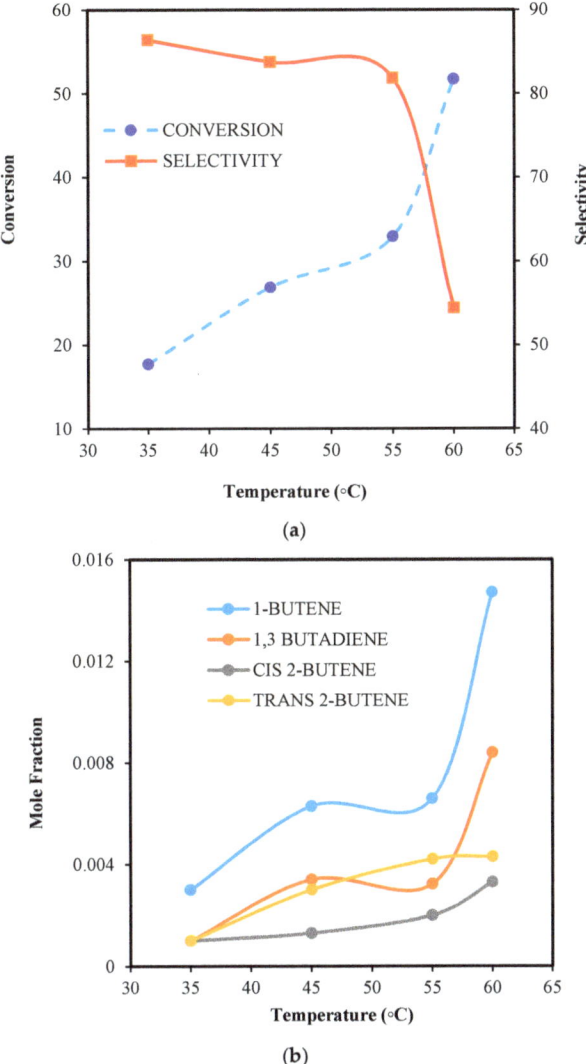

Figure 9. Effect of temperature on (**a**) acetylene conversion, ethylene selectivity, and (**b**) product distribution at 18 bar, hydrogen to acetylene ratio 1.2 and GHSV 5500 1/h.

6.2.4. Effect of Hydrogen to Acetylene Ratio

In general, the presence of excess hydrogen in the reactor, reduces coke build-up on the catalyst, and consequently retards the deactivation of catalysts in the hydrogenation process. Figure 10a,b shows the effect of hydrogen to acetylene ratio on acetylene conversion, ethylene selectivity, and product distribution. Increasing hydrogen concentration in the reactor increases the rate of hydrogenation reactions and results in higher acetylene conversion. Although increasing the hydrogen to acetylene ratio enhances the rate of acetylene hydrogenation, it shifts the ethylene hydrogenation toward higher conversion and decreases process selectivity. It appears that applying hydrogen rich stream increases 1-butene concentration in the reactor.

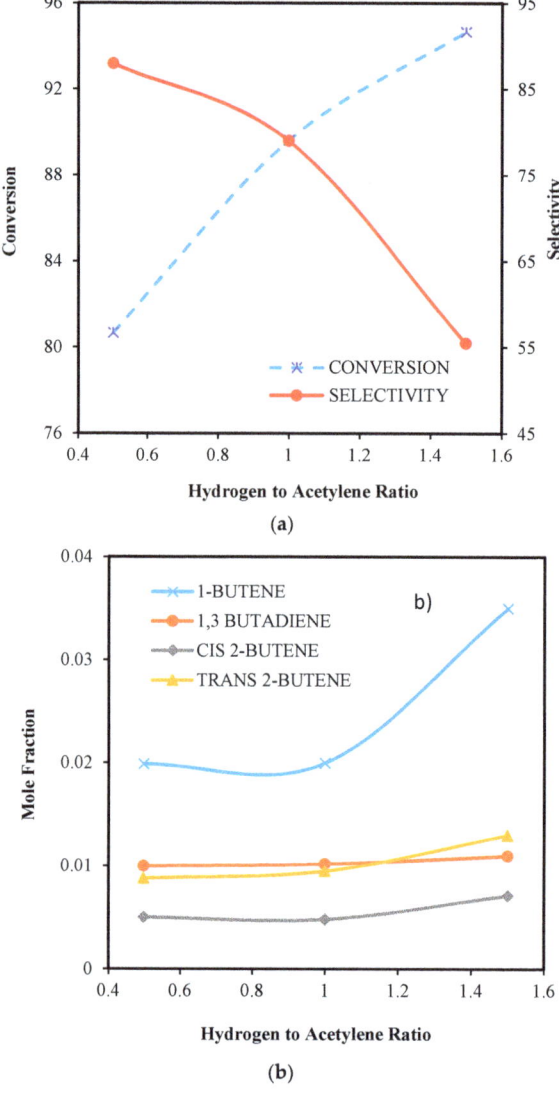

Figure 10. The effect of (**a**) hydrogen to acetylene ratio on acetylene conversion, ethylene selectivity, and (**b**) product distribution at 20 bar, 45 °C and GHSV 2600 1/h.

6.2.5. Developed Kinetic and Decay Models

In this section, the proposed kinetic model, which predict the rate of reactions, is presented. In this regard, an optimization problem is formulated and the coefficient of proposed rate equations are determined by considering the absolute difference between experimental data and estimated rate as the objective function. Table 5 shows the reactions and proposed kinetic equations to calculate the reaction rate.

Table 5. Kinetics of reactions and proposed kinetic model.

Reaction	Proposed Kinetic Model
$C_2H_2 + H_2 \rightarrow C_2H_4$	$r_i = 0.0197 e^{-\frac{11641.3}{RT}} \dfrac{p_{C_2H_2}^{0.4} \times p_{H_2}^{0.9}}{\left(1 + K_{C_2H_2} p_{C_2H_2}\right)^{1.1}}$
$C_2H_4 + H_2 \rightarrow C_2H_6$	$r_i = 0.0098 e^{-\frac{6067.7}{RT}} \dfrac{p_{C_2H_2}^{1.4} \times p_{H_2}^{1.4}}{\left(1 + K_{C_2H_4} p_{C_2H_4}\right)^{1.6}}$
$2\,C_2H_2 + H_2 \rightarrow C_4H_6$	$r_i = 0.0032 e^{-\frac{14174.1}{RT}} \dfrac{p_{C_2H_2}^{0.4} \times p_{H_2}^{1.8}}{\left(1 + K_{C_2H_2} p_{C_2H_2}\right)^{2}}$
$2\,C_2H_2 + 2H_2 \rightarrow C_4H_8$	$r_i = 0.00027 e^{-\frac{21020.1}{RT}} \dfrac{p_{C_2H_2}^{1.1} \times p_{H_2}^{1.7}}{\left(1 + K_{C_2H_2} p_{C_2H_2}\right)^{1.7}}$

Where

$$K_{C_2H_2} = 2.128 \times e^{\frac{2983.8}{RT}} \qquad (27)$$

$$K_{C_2H_4} = 0.7295 \times e^{\frac{3621}{RT}} \qquad (28)$$

The coefficient of the proposed deactivation model is calculated based on the integration of process model and developed the kinetic model. The proposed activity model is inserted in the developed dynamic model of the acetylene hydrogenation process. Then, an optimization problem is formulated to calculate the parameters of the proposed activity model, K_d, E_d, and n, considering the sum of absolute difference between plant data and simulation results, during a process run-time as the objective function. The industrial data points have been presented in the Supplementary Data Set 2. In addition, the composition of green oil as a deactivation agent is presented in the Supplementary Data Set 3. The obtained deactivation model could be explained as:

$$\frac{da}{dt} = -0.21\, e^{-\left(\frac{9504.4}{RT}\right)} \times a^{2.4} \times C^{0.13} \qquad (29)$$

6.3. Results of Process Simulation

In this section, the simulation result presents the accuracy of the developed model and the assumptions are proved at the dynamic condition. Then an optimization problem is formulated and the optimal operating condition of the process is determined to increase process run time.

6.3.1. Model Validation

In this research, two different methods are utilized to investigate the accuracy of the developed kinetic model [31]. Thermodynamically, the comparison adsorption constants of compartments and the thermo-dynamic value, present a quality base criterion to investigate validity of kinetic equation. In this regard, combining the entropy concept with gas universal constant provide a procedure for finding the thermodynamic compatibility. In detail, if the overall entropy of the gaseous state of components is higher than the entropy of adsorbed components, the thermo-dynamic compatibility is reached. In more detail, according to this concept, if the kinematic constants match the following equations, it is possible to claim the thermodynamic compatibility:

$$\Delta_{ads}S^0_{i.j} = S^0_{ads.i.j} - S^0_{g.i} < 0 \tag{30}$$

$$exp\frac{\Delta_{ads}S^0_{i.j}}{R} = K_{j.i\infty} \tag{31}$$

$$\left|\Delta_{ads}S^0_{i.j}\right| < S^0_{g.i} \tag{32}$$

$$\left|\Delta_{ads}S^0_{i.j}\right| > -R.\ln\frac{\vartheta_i}{\vartheta_{cr.i}} \approx 41.8\, J/(mol.K) \tag{33}$$

$$\Delta_{ads}S^0_{i.j} < -51\frac{J}{mol.K} + \frac{0.00141}{K} \times \Delta_{ads}H_{i.j} \tag{34}$$

In addition, to prove the validity of the developed model, the simulation results are compared with the real plant data at the dynamic condition. Figure 11a,b shows the comparison between outlet acetylene concentration from guard bed and calculated concentration by the model. The mean absolute error of the model and plant data is below 3.0%. Thus, the proposed model is a practical tool in predicting the performance of a hydrogenation process.

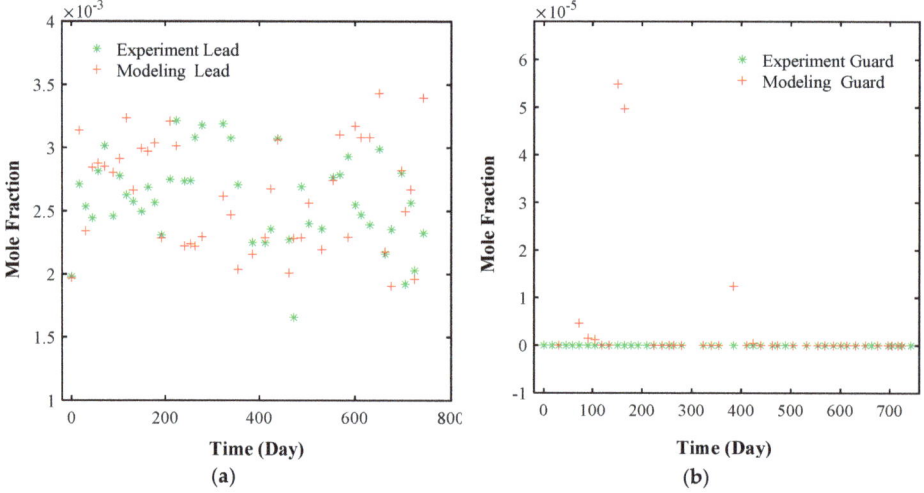

Figure 11. Comparison between outlet acetylene concentrations calculated by the model and plant data in (**a**) lead and, (**b**) guard beds.

6.3.2. Reactor Simulation

In this section, the concentration and temperature profiles, along the reactors, are presented during the process run-time. Based on the simulation results, after 400 days of continuous operation, the activity of catalyst in the Lead bed decreased to 0.2, while the activity of the catalyst in the Guard bed is 0.5. Figure 12 shows the acetylene molar flow rate along the Lead and guard beds during the process run-time. It appears that the acetylene concentration decreases along the reactor length. Due to catalyst deactivation, the acetylene concentration in the outlet stream from lead bed increases during the process run-time and approaches from 7.43 mol s^{-1} to 10.09 mol s^{-1}. Typically, the acetylene conversion decreases during the process run-time in the Lead bed and approaches from 67.2% at the start of the run to 55.5% at the end of run. Decreasing acetylene conversion in the Lead bed proves the philosophy of the Guard bed in the acetylene hydrogenation process. The unconverted acetylene is converted to ethane and ethylene in the Guard bed. It appears that acetylene molar flow rate in the outlet stream from the Guard bed increases during the process run-time and approaches from 0.19 to 0.21 mol s^{-1}.

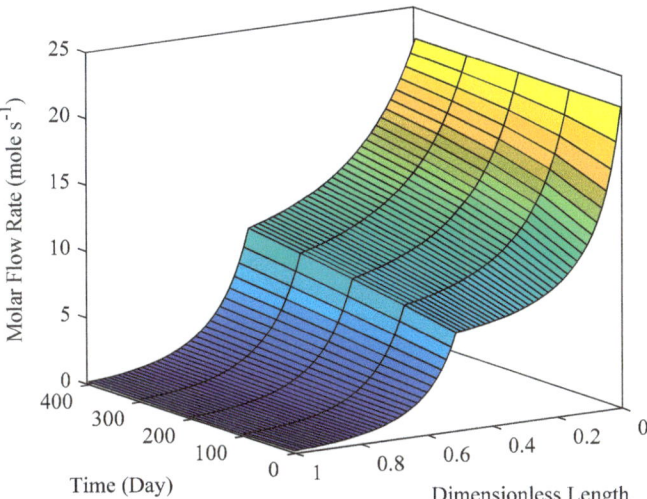

Figure 12. Acetylene flow rate along the Lead and guard beds during the process runtime.

Figure 13 shows the temperature profile along the Lead and Guard beds during the process runtime. Since the acetylene hydrogenation reaction is exothermic, temperature increases along the reactors. Typically, catalyst decay decreases the rate of acetylene hydrogenation in the Lead and guard beds, and the temperature of outlet stream from the Lead and guard beds decreases gradually. Lower acetylene hydrogenation in the Lead and guard beds increases acetylene concentration in the feed of Guard bed reactor during the process run time. Thus, increasing acetylene concentration in the Guard bed increases heat generation through a hydrogenation reaction and temperature increases at the outlet of Guard bed. Generally, lower acetylene conversion in the Guard bed results in the lower temperate rise in the reactor.

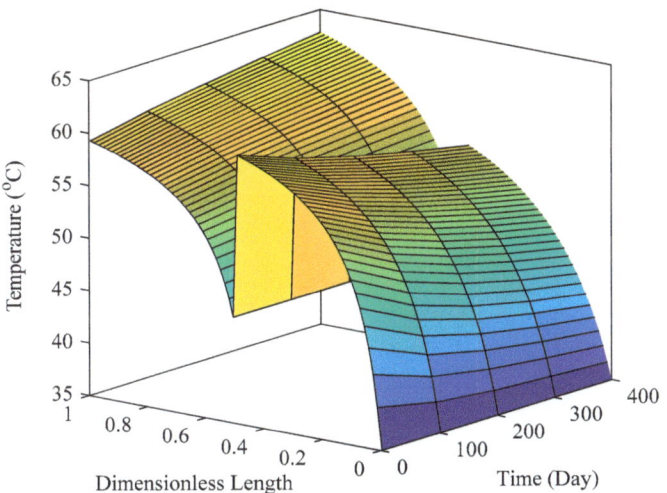

Figure 13. Temperature profile along the Lead and Guard beds during the process runtime.

7. Conclusions

In this research, the acetylene hydrogenation over Pd-Ag supported α-Al_2O_3 was investigated in a differential reactor. The full factorial design method, based on the cubic pattern was used to determine the experiments, considering hydrogen to acetylene ratio, temperature, gas hourly space velocity, and pressure as dependent variables. The fresh and spent catalysts were characterized by SEM, TEM, DTG-TGA, XRD, and BET tests. It is concluded from SEM and BET tests, that coke build-up on the external surface of the catalyst increases surface area and decrease pore mean diameter. Then, a detailed reaction network was proposed based on the Langmuir-Hinshelwood-Hougen-Watson approach, considering ethane, 1-butene, and 1,3-butadine as side products. The coefficients of the proposed kinetic model were calculated, based on experimental data. In addition, the industrial Tail-End hydrogenation reactors were modelled, and a decay model was proposed to predict catalyst activity. The results showed that applying hydrogen rich stream increases 1-butene concentration in the reactor. Based on the simulation results the acetylene molar flow rate in the outlet stream from Guard bed increases during the process run-time and approaches from 0.19 to 0.21 mol s^{-1}.

Supplementary Materials: The following are available online at http://www.mdpi.com/2227-9717/7/3/136/s1.

Author Contributions: O.D.: Conceived and designed the analysis, Collected the data, Contributed data or analysis tools, Performed the analysis, Wrote the paper; M.R.R.: Conceived and designed the analysis, Contributed data or analysis tools, Performed the analysis, Wrote the paper; A.S.: Conceived and designed the analysis, Performed the analysis, Wrote the paper.

Conflicts of Interest: The authors confirm that there are no known conflict of interest associated with this publication and there has been no significant financial support for this work that could have influenced its outcome.

Nomenclature

ΔH	enthalpy of reaction
MFC	mass flow control
TIC	temperature indicator controller
PIC	pressure indicator controller
FIC	flow indicator controller
Re	Reynolds number
L	reactor length
D	reactor diameter
d_p	catalyst diameter
Q_g	gas flow rate (experimental)
GHSV	gas hourly space velocity
r_i	overall rate of reaction
k	constant of reaction
K	constant of adsorption
P	pressure
n	power of reaction rate nominator
α	power of reaction rate denominator
A_0	Arrenius type constant
A_{ij}	constant of adsorption
R	gas constant
R_j	local (component) rate of reaction
V_{cat}	volume of catalyst
a	activity of catalyst
T	temperature
t	time
E	activation energy
E_d	activation energy of deactivation equation
k_d	constant of deactivation equation
GC	gas chromatography

m_i	power of nominator and denominator of reaction rate equation (i = 1–12)
n_i	power of nominator of deactivation equation (i = 1–2)
MW	molecular weight
TC	critical temperature
PC	critical pressure
C_p	heat capacity
A,B,C,D	constant of heat capacity equation
μcr	critical viscosity
μ	viscosity
A	surface area
Z	length
NA	mole flux
ρb	bulk density
ε	porosity
MRE	mean relative error (N: number of component), (exp: experiment)
Z	z factor
S	vacant site
s	Entropy

References

1. Adams, D.; Blankenship, S.; Geyer, I.; Takenaka, T. Front end & back end acetylene converter catalysts. In Proceedings of the 3rd Asian Ethylene Symposium on Catalyst and Processes, Yokohama, Japan, 4–6 October 2000; pp. 4–6.
2. Barazandeh, K.; Dehghani, O.; Hamidi, M.; Aryafard, E.; Rahimpour, M.R. Investigation of coil outlet temperature effect on the performance of naphtha cracking furnace. *Chem. Eng. Res. Des.* **2015**, *94*, 307–316. [CrossRef]
3. Huang, W.; McCormick, J.R.; Lobo, R.F.; Chen, J.G. Selective hydrogenation of acetylene in the presence of ethylene on zeolite-supported bimetallic catalysts. *J. Catal.* **2007**, *246*, 40–51. [CrossRef]
4. Miller, S.A. *Acetylene: Its Properties, Manufacture, and Uses*; Academic Press: Cambridge, MA, USA, 1966; Volume 2.
5. Kadiva, A.; Sadeghi, M.T.; Sotudeh-Gharebagh, R.; Mahmudi, M. Estimation of kinetic parameters for hydrogenation reactions using a genetic algorithm. *Chem. Eng. Technol.* **2009**, *32*, 1588–1594. [CrossRef]
6. Mansoornejad, B.; Mostoufi, N.; Jalali-Farahani, F. A hybrid GA–SQP optimization technique for determination of kinetic parameters of hydrogenation reactions. *Comput. Chem. Eng.* **2008**, *32*, 1447–1455. [CrossRef]
7. Mostoufi, N.; Ghoorchian, A.; Sotudeh-Gharebagh, R. Hydrogenation of acetylene: Kinetic studies and reactor modeling. *In. J. Chem. React. Eng.* **2005**, *3*. [CrossRef]
8. Ravanchi, M.T.; Sahebdelfar, S.; Komeili, S. Acetylene selective hydrogenation: A technical review on catalytic aspects. *Rev. Chem. Eng.* **2018**, *34*, 215–237. [CrossRef]
9. Bos, A.; Botsma, E.; Foeth, F.; Sleyster, H.; Westerterp, K. A kinetic study of the hydrogenation of ethyne and ethene on a commercial Pd/Al$_2$O$_3$ catalyst. *Chem. Eng. Process. Process Intensif.* **1993**, *32*, 53–63. [CrossRef]
10. Borodziński, A. Hydrogenation of acetylene–ethylene mixtures on a commercial palladium catalyst. *Catal. Lett.* **1999**, *63*, 35–42. [CrossRef]
11. Zhang, Q.; Li, J.; Liu, X.; Zhu, Q. Synergetic effect of pd and ag dispersed on Al$_2$O$_3$ in the selective hydrogenation of acetylene. *Appl. Catal. A Gen.* **2000**, *197*, 221–228. [CrossRef]
12. Sarkany, A.; Horvath, A.; Beck, A. Hydrogenation of acetylene over low loaded pd and pd-au/SiO$_2$ catalysts. *Appl. Catal. A Gen.* **2002**, *229*, 117–125. [CrossRef]
13. Schbib, N.S.; García, M.A.; Gígola, C.E.; Errazu, A.F. Kinetics of front-end acetylene hydrogenation in ethylene production. *Ind. Eng. Chem. Res.* **1996**, *35*, 1496–1505. [CrossRef]
14. Flick, K.; Herion, C.; Allmann, H.-M. Supported Palladium Catalyst for Selective Catalytic Hydrogenation of Acetylene in Hydrocarbonaceous Streams. U.S. Patent US5856262A, 5 January 1999.
15. Khan, N.A.; Shaikhutdinov, S.; Freund, H.-J. Acetylene and ethylene hydrogenation on alumina supported pd-ag model catalysts. *Catal. Lett.* **2006**, *108*, 159–164. [CrossRef]

16. Aduriz, H.; Bodnariuk, P.; Dennehy, M.; Gigola, C. Activity and selectivity of Pd/α-Al$_2$O$_3$ for ethyne hydrogenation in a large excess of ethene and hydrogen. *Appl. Catal.* **1990**, *58*, 227–239. [CrossRef]
17. Pachulski, A.; Schödel, R.; Claus, P. Performance and regeneration studies of Pd–Ag/Al$_2$O$_3$ catalysts for the selective hydrogenation of acetylene. *Appl. Catal. A Gen.* **2011**, *400*, 14–24. [CrossRef]
18. Tejeda-Serrano, M.A.; Mon, M.; Ross, B.; Gonell, F.; Ferrando-Soria, J.S.; Corma, A.; Leyva-Pérez, A.; Armentano, D.; Pardo, E. Isolated Fe (iii)–o sites catalyze the hydrogenation of acetylene in ethylene flows under front-end industrial conditions. *J. Am. Chem. Soc.* **2018**, *140*, 8827–8832. [CrossRef] [PubMed]
19. Gobbo, R.; Soares, R.d.P.; Lansarin, M.A.; Secchi, A.R.; Ferreira, J.M.P. Modeling, simulation, and optimization of a front-end system for acetylene hydrogenation reactors. *Braz. J. Chem. Eng.* **2004**, *21*, 545–556. [CrossRef]
20. Aeowjaroenlap, H.; Chotiwiriyakun, K.; Tiensai, N.; Tanthapanichakoon, W.; Spatenka, S.; Cano, A. Model-based optimization of an acetylene hydrogenation reactor to improve overall ethylene plant economics. *Ind. Eng. Chem. Res.* **2018**, *57*, 9943–9951. [CrossRef]
21. Samavati, M.; Ebrahim, H.A.; Dorj, Y. Effect of the operating parameters on the simulation of acetylene hydrogenation reactor with catalyst deactivation. *Appl. Catal. A Gen.* **2018**, *567*, 45–55. [CrossRef]
22. Khold, O.D.; Parhoudeh, M.; Rahimpour, M.R.; Raeissi, S. A new configuration in the tail-end acetylene hydrogenation reactor to enhance catalyst lifetime and performance. *J. Taiwan Inst. Chem. Eng.* **2016**, *65*, 8–21. [CrossRef]
23. Kirk, R.E. Experimental design. In *The Blackwell Encyclopedia of Sociology*; Wiley: Hoboken, NJ, USA, 2007.
24. Bos, A.; Westerterp, K. Mechanism and kinetics of the selective hydrogenation of ethyne and ethene. *Chem. Eng. Process. Process Intensif.* **1993**, *32*, 1–7. [CrossRef]
25. Oudar, J. *Deactivation and Poisoning of Catalysts*; CRC Press: Boca Raton, FL, USA, 1985; Volume 20.
26. Fogler, H.S. *Elements of Chemical Reaction Engineering*; Prentice Hall: Upper Sanddle River, NJ, USA, 1999.
27. Bartholomew, C. Catalyst deactivation and regeneration. In *Kirk-Othmer Encyclopedia of Chemical Technology*; Wiley: Hoboken, NJ, USA, 2000.
28. Sarkany, A.; Guczi, L.; Weiss, A.H. On the aging phenomenon in palladium catalysed acetylene hydrogenation. *Appl. Catal.* **1984**, *10*, 369–388. [CrossRef]
29. Robinson, D. Catalyst regeneration, metal catalysts. In *Kirk-Othmer Encyclopedia of Chemical Technology*; Wiley: Hoboken, NJ, USA, 2000.
30. Anderson-Cook, C.M. *Practical Genetic Algorithms*; Taylor & Francis: Abingdon, UK, 2005.
31. Boudart, M.; Djéga-Mariadassou, G. *Kinetics of Heterogeneous Catalytic Reactions*; Princeton University Press: Princeton, NJ, USA, 2014; Volume 767.

© 2019 by the authors. Licensee MDPI, Basel, Switzerland. This article is an open access article distributed under the terms and conditions of the Creative Commons Attribution (CC BY) license (http://creativecommons.org/licenses/by/4.0/).

MDPI
St. Alban-Anlage 66
4052 Basel
Switzerland
Tel. +41 61 683 77 34
Fax +41 61 302 89 18
www.mdpi.com

Processes Editorial Office
E-mail: processes@mdpi.com
www.mdpi.com/journal/processes

www.ingramcontent.com/pod-product-compliance
Lightning Source LLC
LaVergne TN
LVHW070557100526
838202LV00012B/496